Earth and Planetary Science

Edited by
Harv Kennedy

Larsen & Keller
www.larsen-keller.com

Earth and Planetary Science
Edited by Harv Kennedy
ISBN: 978-1-63549-089-3 (Hardback)

© 2017 Larsen & Keller

☰ Larsen & Keller

Published by Larsen and Keller Education,
5 Penn Plaza,
19th Floor,
New York, NY 10001, USA

Cataloging-in-Publication Data

Earth and planetary science / edited by Harv Kennedy.
    p. cm.
Includes bibliographical references and index.
ISBN 978-1-63549-089-3
1. Earth sciences. 2. Planetary science.
I. Kennedy, Harv.
QE28.2 .E27 2017
550--dc23

The publisher's policy is to use permanent paper from mills that operate a sustainable forestry policy. Furthermore, the publisher ensures that the text paper and cover boards used have met acceptable environmental accreditation standards.

Printed and bound in the United States of America.

For more information regarding Larsen and Keller Education and its products, please visit the publisher's website www.larsen-keller.com

Table of Contents

Preface

Planetary sciences is a vast field which studies the planets, planetary systems and natural satellites. Earth sciences incorporates the rules and laws of physics, mathematics, biology and chemistry, etc. to study the various resources and elements found on earth. This book attempts to understand the multiple branches that fall under the discipline of Earth and planetary sciences and how such concepts have practical applications. Such selected concepts that redefine this subject have been presented in the text. This book presents the complex subject of Earth and planetary science in the most comprehensible and easy to understand language. It will serve as a valuable source of reference for those interested in this field.

A detailed account of the significant topics covered in this book is provided below:

Chapter 1- The planet Earth has a unique history among the known celestial bodies as it has sustained life. The character of its atmosphere and geosphere has enabled this to happen. Many other planets have similar as well as differing characteristics and elements. This chapter is an overview of the subject matter incorporating all the major aspects of Earth Science and Planetary Science.

Chapter 2- The planetary functions of the earth can be divided into many spheres. They have been divided for detailed scientific study. Some of these are the atmosphere, geosphere, biosphere etc. They provide vital energy to each other as well as enable life on earth. The major components of the Earth are discussed in this chapter.

Chapter 3- Study of the Earth and its various functions such as the change of seasons, tidal movements and seismic activity give us an idea of the structure and nature of life in other planets as well as our own in different eras. This chapter is a compilation of the various branches of Earth and Planetary Science that form an integral part of the broader subject matter.

Chapter 4- Study of the major activities of the spheres that comprise the Earth's functions enable the study and understanding of spatial and temporal changes in climate, atmospheric pressure, sea-water levels and many other events. They help to correlate weather patterns as well as other instances of planetary activity and their mutual influence on each other.

Chapter 5- The Earth is not a monolith and it has gone through many changes in its atmosphere, climate, as well as geological structures. These changes have been faithfully preserved in geologic and other fossilized records that are available to scientists. They help to understand and predict changes in climate, seismic activities and volcanic activities of the Earth. The topics discussed in the chapter are of great importance to broaden the existing knowledge on earth and planetary science.

Chapter 6- Earth system science offers a holistic view into the functioning of planet Earth. It considers the movement of the planet along with the changes in the seasons as subjects that need to be studied together rather than separately. It also considers human activity such as waste disposal and emissions to be objects of study along with celestial activity.

It gives me an immense pleasure to thank our entire team for their efforts. Finally in the end, I would like to thank my family and colleagues who have been a great source of inspiration and support.

Editor

Introduction to Earth and Planetary Science

The planet Earth has a unique history among the known celestial bodies as it has sustained life. The character of its atmosphere and geosphere has enabled this to happen. Many other planets have similar as well as differing characteristics and elements. This chapter is an overview of the subject matter incorporating all the major aspects of Earth Science and Planetary Science.

Earth Science

Earth science or geoscience is an all-encompassing term that refers to the fields of science dealing with Planet Earth. It can be considered to be a branch of planetary science, but with a much older history. There are both reductionist and holistic approaches to Earth sciences. The formal discipline of Earth sciences may include the study of the atmosphere, hydrosphere, lithosphere, and biosphere. Typically, Earth scientists will use tools from physics, chemistry, biology, chronology, and mathematics to build a quantitative understanding of how Earth system works and how it evolved to its current state.

Fields of Study

The following fields of science are generally categorized within the Earth sciences:

- Geography specifically Physical Geography covers aspects of geomorphology, soil study, hydrology, meteorology, climatology, and biogeography.

- Geology describes the rocky parts of the Earth's crust (or lithosphere) and its historic development. Major subdisciplines are mineralogy and petrology, geochemistry, geomorphology, paleontology, stratigraphy, structural geology, engineering geology, and sedimentology.

- Geophysics and geodesy investigate the shape of the Earth, its reaction to forces and its magnetic and gravity fields. Geophysicists explore the Earth's core and mantle as well as the tectonic and seismic activity of the lithosphere. Geophysics is commonly used to supplement the work of geologists in developing a comprehensive understanding of crustal geology, particularly in mineral and petroleum exploration.

- Soil science covers the outermost layer of the Earth's crust that is subject to soil formation processes (or pedosphere). Major subdisciplines include edaphology and pedology.

- Ecology covers the interactions between the biota, with their natural environment. This field of study differentiates the study of the Earth, from the study of other planets in the Solar System; the Earth being the only planet teeming with life.

- Hydrology (includes oceanography and limnology) describe the marine and freshwater domains of the watery parts of the Earth (or hydrosphere). Major subdisciplines include hydrogeology and physical, chemical, and biological oceanography.

- Glaciology covers the icy parts of the Earth (or cryosphere).

- Atmospheric sciences cover the gaseous parts of the Earth (or atmosphere) between the surface and the exosphere (about 1000 km). Major subdisciplines include meteorology, climatology, atmospheric chemistry, and atmospheric physics.

Earth's Interior

A volcanic eruption is the release of stored energy from below the surface of Earth.

Plate tectonics, mountain ranges, volcanoes, and earthquakes are geological phenomena that can be explained in terms of energy transformations in Earth's crust.

Beneath Earth's crust lies the mantle which is heated by the radioactive decay of heavy elements. The mantle is not quite solid and consists of magma which is in a state of semi-perpetual convection. This convection process causes the lithospheric plates to move, albeit slowly. The resulting process is known as plate tectonics.

Plate tectonics might be thought of as the process by which the earth is resurfaced. Through a process called seafloor spreading, new crust is created by the flow of magma from underneath the lithosphere to the surface, through fissures, where it cools and solidifies. Through a process called subduction, oceanic crust is pushed underground — beneath the rest of the lithosphere—where it comes into contact with magma and melts—rejoining the mantle from which it originally came.

Areas of the crust where new crust is created are called divergent boundaries, those where it is brought back into the earth are convergent boundaries and those where plates slide past each other, but no new lithospheric material is created or destroyed, are referred to as transform (or conservative) boundaries Earthquakes result from the movement of the lithospheric plates, and they often occur near convergent boundaries where parts of the crust are forced into the earth as part of subduction.

Volcanoes result primarily from the melting of subducted crust material. Crust material that is forced into the asthenosphere melts, and some portion of the melted material becomes light enough to rise to the surface—giving birth to volcanoes.

Earth's Electromagnetic Field

An electromagnet is a magnet that is created by a current that flows around a soft iron core. Earth has a solid iron inner core surrounded by semi-liquid materials of the outer core that move in continuous currents around the inner core; therefore, the Earth is an electromagnet. This is referred to as the dynamo theory of Earth's magnetism.

Earth's Atmosphere

The magnetosphere shields the surface of Earth from the charged particles of the solar wind. (*image not to scale.*)

The troposphere, stratosphere, mesosphere, thermosphere, and exosphere are the five layers which make up Earth's atmosphere. In all, the atmosphere is made up of about 78.0% nitrogen, 20.9% oxygen, and 0.92% argon. 75% of the gases in the atmosphere are located within the troposphere, the bottom-most layer. The remaining one percent of the atmosphere (all but the nitrogen, oxygen, and argon) contains small amounts of other gases including CO_2 and water vapors. Water vapors and CO_2 allow the Earth's atmosphere to catch and hold the Sun's energy through a phenomenon called the greenhouse effect. This allows Earth's surface to be warm enough to have liquid water and support life.

The magnetic field created by the internal motions of the core produces the magnetosphere which protects the Earth's atmosphere from the solar wind. As the earth is 4.5 billion years old, it would have lost its atmosphere by now if there were no protective magnetosphere.

In addition to storing heat, the atmosphere also protects living organisms by shielding the Earth's surface from cosmic rays. Note that the level of protection is high enough to prevent cosmic rays from destroying all life on Earth, yet low enough to aid the mutations that have an important role in pushing forward diversity in the biosphere.

Methodology

Methodologies vary depending on the nature of the subjects being studied. Studies typically fall into one of three categories: observational, experimental, or theoretical. Earth scientists often conduct sophisticated computer analysis or go to many of the world's most exotic locations to study Earth phenomena (e.g. Antarctica or hot spot island chains).

A foundational idea within the study Earth science is the notion of uniformitarianism. Uniformitarianism dictates that "ancient geologic features are interpreted by understanding active processes that are readily observed." In other words, any geologic processes at work in the present have operated in the same ways throughout geologic time. This enables those who study Earth's history to apply knowledge of how Earth processes operate in the present to gain insight into how the planet has evolved and changed throughout deep history.

Earth's Spheres

Earth science generally recognizes four spheres, the lithosphere, the hydrosphere, the atmosphere, and the biosphere; these correspond to rocks, water, air and life. Also included by some are the cryosphere (corresponding to ice) as a distinct portion of the hydrosphere and the pedosphere (corresponding to soil) as an active and intermixed sphere.

Partial list of the major earth science topics

Atmosphere

- Atmospheric chemistry
- Geography
    o Climatology
    o Meteorology
- Hydrometeorology
- Paleoclimatology

Biosphere

- Biogeochemistry
- Biogeography
- Ecology
    o Landscape ecology
- Geoarchaeology
- Geomicrobiology
- Paleontology
    o Palynology
    o Micropaleontology

Hydrosphere

- Hydrology
  - Hydrogeology
- Limnology (freshwater science)
- Oceanography (marine science)
  - Chemical oceanography
  - Physical oceanography
  - Biological oceanography (marine biology)
  - Geological oceanography (marine geology)
    - Paleoceanography

Lithosphere (Geosphere)

- Geology
  - Economic geology
  - Engineering geology
  - Environmental geology
  - Historical geology
    - Quaternary geology
  - Planetary geology and planetary geography
  - Sedimentology
  - Stratigraphy
  - Structural geology
- Geography
  - Physical geography
- Geochemistry
- Geomorphology
- Geophysics
  - Geochronology
  - Geodynamics

- o   Geomagnetism
- o   Gravimetry (also part of Geodesy)
- o   Seismology
- Glaciology
- Hydrogeology
- Mineralogy
  - o   Crystallography
  - o   Gemology
- Petrology
- Speleology
- Volcanology

Pedosphere

- Geography
  - o   Soil science
    - ▪   Edaphology
    - ▪   Pedology

Systems

- Earth system science
- Environmental science
- Geography
  - o   Human geography
  - o   Physical geography
- Gaia hypothesis
- Systems ecology
- Systems geology

Others

- Geography

- o Cartography

- o Geoinformatics (GIS)

- o Geostatistics

- o Geodesy and Surveying

- o NASA Earth Science Enterprise

- o Remote Sensing

Planetary Science

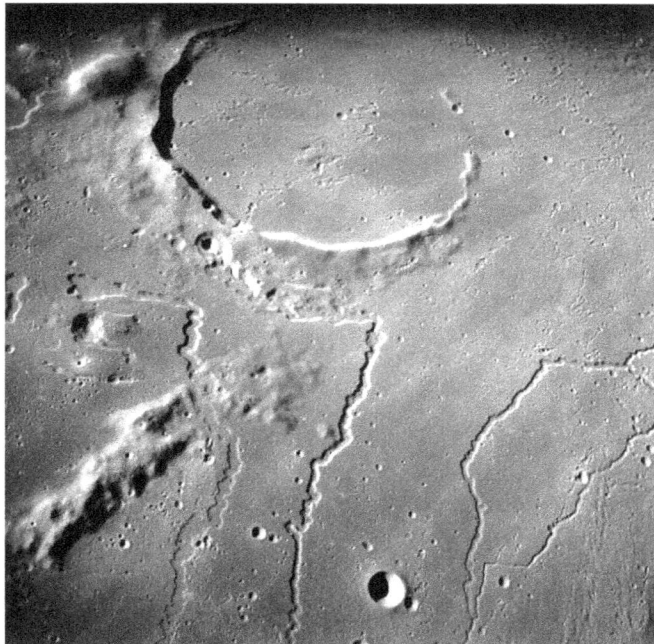

Photograph from Apollo 15 orbital unit of the rilles in the vicinity of the crater Aristarchus on the Moon. The arrangement of the two valleys is very similar, although one third the size, to Great Hungarian Plain rivers Danube and Tisza.

Planetary science or, more rarely, planetology, is the scientific study of planets (including Earth), moons, and planetary systems (in particular those of the Solar System) and the processes that form them. It studies objects ranging in size from micrometeoroids to gas giants, aiming to determine their composition, dynamics, formation, interrelations and history. It is a strongly interdisciplinary field, originally growing from astronomy and earth science, but which now incorporates many disciplines, including planetary geology (together with geochemistry and geophysics), cosmochemistry, atmospheric science, oceanography, hydrology, theoretical planetary science, glaciology, and exoplanetology. Allied disciplines include space physics, when concerned with the effects of the Sun on the bodies of the Solar System, and astrobiology.

There are interrelated observational and theoretical branches of planetary science. Observational research can involve a combination of space exploration, predominantly with robotic spacecraft missions using remote sensing, and comparative, experimental work in Earth-based laboratories. The theoretical component involves considerable computer simulation and mathematical modelling.

Planetary scientists are generally located in the astronomy and physics or Earth sciences departments of universities or research centres, though there are several purely planetary science institutes worldwide. There are several major conferences each year, and a wide range of peer-reviewed journals.

History

The history of planetary science may be said to have begun with the Ancient Greek philosopher Democritus, who is reported by Hippolytus as saying

The ordered worlds are boundless and differ in size, and that in some there is neither sun nor moon, but that in others, both are greater than with us, and yet with others more in number. And that the intervals between the ordered worlds are unequal, here more and there less, and that some increase, others flourish and others decay, and here they come into being and there they are eclipsed. But that they are destroyed by colliding with one another. And that some ordered worlds are bare of animals and plants and all water.

In more modern times, planetary science began in astronomy, from studies of the unresolved planets. In this sense, the original planetary astronomer would be Galileo, who discovered the four largest moons of Jupiter, the mountains on the Moon, and first observed the rings of Saturn, all objects of intense later study. Galileo's study of the lunar mountains in 1609 also began the study of extraterrestrial landscapes: his observation "that the Moon certainly does not possess a smooth and polished surface" suggested that it and other worlds might appear "just like the face of the Earth itself".

Advances in telescope construction and instrumental resolution gradually allowed increased identification of the atmospheric and surface details of the planets. The Moon was initially the most heavily studied, as it always exhibited details on its surface, due to its proximity to the Earth, and the technological improvements gradually produced more detailed lunar geological knowledge. In this scientific process, the main instruments were astronomical optical telescopes (and later radio telescopes) and finally robotic exploratory spacecraft.

The Solar System has now been relatively well-studied, and a good overall understanding of the formation and evolution of this planetary system exists. However, there are large numbers of unsolved questions, and the rate of new discoveries is very high, partly due to the large number of interplanetary spacecraft currently exploring the Solar System.

Disciplines

Planetary Astronomy

This is both an observational and a theoretical science. Observational researchers are predominantly concerned with the study of the small bodies of the Solar System: those that are observed by telescopes, both optical and radio, so that characteristics of these bodies such as shape, spin, surface materials and weathering are determined, and the history of their formation and evolution can be understood.

Theoretical planetary astronomy is concerned with dynamics: the application of the principles of celestial mechanics to the Solar System and extrasolar planetary systems.

Planetary Geology

The best known research topics of planetary geology deal with the planetary bodies in the near vicinity of the Earth: the Moon, and the two neighbouring planets: Venus and Mars. Of these, the Moon was studied first, using methods developed earlier on the Earth.

Geomorphology

Geomorphology studies the features on planetary surfaces and reconstructs the history of their formation, inferring the physical processes that acted on the surface. Planetary geomorphology includes study of several classes of surface feature:

- Impact features (multi-ringed basins, craters)

- Volcanic and tectonic features (lava flows, fissures, rilles)

- Space weathering - erosional effects generated by the harsh environment of space (continuous micrometeorite bombardment, high-energy particle rain, impact gardening). For example, the thin dust cover on the surface of the lunar regolith is a result of micrometeorite bombardment.

- Hydrological features: the liquid involved can range from water to hydrocarbon and ammonia, depending on the location within the Solar System.

The history of a planetary surface can be deciphered by mapping features from top to bottom according to their deposition sequence, as first determined on terrestrial strata by Nicolas Steno. For example, stratigraphic mapping prepared the Apollo astronauts for the field geology they would encounter on their lunar missions. Overlapping sequences were identified on images taken by the Lunar Orbiter program, and these were used to prepare a lunar stratigraphic column and geological map of the Moon.

Cosmochemistry, Geochemistry and Petrology

One of the main problems when generating hypotheses on the formation and evolution of objects in the Solar System is the lack of samples that can be analysed in the laboratory, where a large suite of tools are available and the full body of knowledge derived from terrestrial geology can be brought to bear. Fortunately, direct samples from the Moon, asteroids and Mars are present on Earth, removed from their parent bodies and delivered as meteorites. Some of these have suffered contamination from the oxidising effect of Earth's atmosphere and the infiltration of the biosphere, but those meteorites collected in the last few decades from Antarctica are almost entirely pristine.

The different types of meteorite that originate from the asteroid belt cover almost all parts of the structure of differentiated bodies: meteorites even exist that come from the core-mantle boundary (pallasites). The combination of geochemistry and observational astronomy has also made it possible to trace the HED meteorites back to a specific asteroid in the main belt, 4 Vesta.

The comparatively few known Martian meteorites have provided insight into the geochemical composition of the Martian crust, although the unavoidable lack of information about their points of origin on the diverse Martian surface has meant that they do not provide more detailed constraints on theories of the evolution of the Martian lithosphere. As of July 24, 2013 65 samples of Martian meteorites have been discovered on Earth. Many were found in either Antarctica or the Sahara Desert.

During the Apollo era, in the Apollo program, 384 kilograms of lunar samples were collected and transported to the Earth, and 3 Soviet Luna robots also delivered regolith samples from the Moon. These samples provide the most comprehensive record of the composition of any Solar System body beside the Earth. The numbers of lunar meteorites are growing quickly in the last few years – as of April 2008 there are 54 meteorites that have been officially classified as lunar. Eleven of these are from the US Antarctic meteorite collection, 6 are from the Japanese Antarctic meteorite collection, and the other 37 are from hot desert localities in Africa, Australia, and the Middle East. The total mass of recognized lunar meteorites is close to 50 kg.

Geophysics

Space probes made it possible to collect data in not only the visible light region, but in other areas of the electromagnetic spectrum. The planets can be characterized by their force fields: gravity and their magnetic fields, which are studied through geophysics and space physics.

Measuring the changes in acceleration experienced by spacecraft as they orbit has allowed fine details of the gravity fields of the planets to be mapped. For example, in the 1970s, the gravity field disturbances above lunar maria were measured through lunar orbiters, which led to the discovery of concentrations of mass, mascons, beneath the Imbrium, Serenitatis, Crisium, Nectaris and Humorum basins.

If a planet's magnetic field is sufficiently strong, its interaction with the solar wind forms a magnetosphere around a planet. Early space probes discovered the gross dimensions of the terrestrial magnetic field, which extends about 10 Earth radii towards the Sun. The solar wind, a stream of charged particles, streams out and around the terrestrial magnetic field, and continues behind the magnetic tail, hundreds of Earth radii downstream. Inside the magnetosphere, there are relatively dense regions of solar wind particles, the Van Allen radiation belts.

Geophysics includes seismology and tectonophysics, geophysical fluid dynamics, mineral physics, geodynamics, mathematical geophysics, and geophysical surveying.

Geodesy, also called geodetics, deals with the measurement and representation of the planets of the Solar System, their gravitational fields and geodynamic phenomena (polar motion in three-dimensional, time-varying space. The science of geodesy has elements of both astrophysics and planetary sciences. The shape of the Earth is to a large extent the result of its rotation, which causes its equatorial bulge, and the competition of geologic processes such as the collision of plates and of vulcanism, resisted by the Earth's gravity field. These principles can be applied to the solid surface of Earth (orogeny; Few mountains are higher than 10 km (6 mi), few deep sea trenches deeper than that because quite simply, a mountain as tall as, for example, 15 km (9 mi), would develop so much pressure at its base, due to gravity, that the rock there would become plastic, and the mountain would slump back to a height of roughly 10 km (6 mi) in a geologically insignificant time. Some or all of these geologic principles can be applied to other planets besides Earth. For instance on Mars, whose surface gravity is much less, the largest volcano, Olympus Mons, is 27 km (17 mi) high at its peak, a height that could not be maintained on Earth. The Earth geoid is essentially the figure of the Earth abstracted from its topographic features. Therefore, the Mars geoid is essentially the figure of Mars abstracted from its topographic features. Surveying and mapping are two important fields of application of geodesy.

Atmospheric Science

Cloud bands clearly visible on Jupiter.

The atmosphere is an important transitional zone between the solid planetary surface and the higher rarefied ionizing and radiation belts. Not all planets have atmospheres: their existence depends on the mass of the planet, and the planet's distance from the Sun — too distant and frozen atmospheres occur. Besides the four gas giant planets, almost all of the terrestrial planets (Earth, Venus, and Mars) have significant atmospheres. Two moons have significant atmospheres: Saturn's moon Titan and Neptune's moon Triton. A tenuous atmosphere exists around Mercury.

The effects of the rotation rate of a planet about its axis can be seen in atmospheric streams and currents. Seen from space, these features show as bands and eddies in the cloud system, and are particularly visible on Jupiter and Saturn.

Comparative Planetary Science

Planetary science frequently makes use of the method of comparison to give greater understanding of the object of study. This can involve comparing the dense atmospheres of Earth and Saturn's moon Titan, the evolution of outer Solar System objects at different distances from the Sun, or the geomorphology of the surfaces of the terrestrial planets, to give only a few examples.

The main comparison that can be made is to features on the Earth, as it is much more accessible and allows a much greater range of measurements to be made. Earth analogue studies are particularly common in planetary geology, geomorphology, and also in atmospheric science.

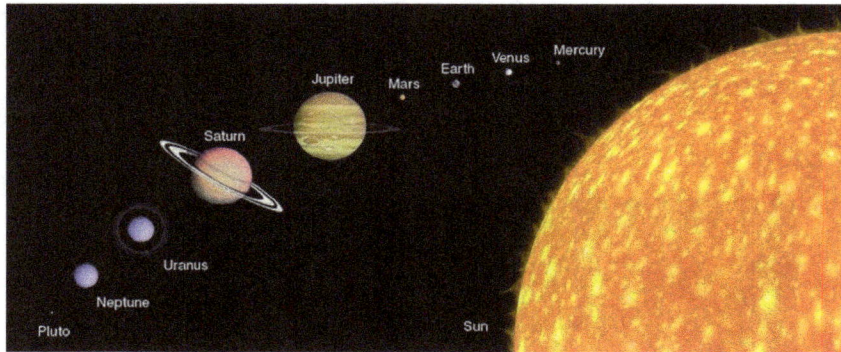

Professional Activity

Professional Bodies

- Division for Planetary Sciences (DPS) of the American Astronomical Society

- American Geophysical Union

- Meteoritical Society

- Europlanet

Major Conferences

- Lunar and Planetary Science Conference (LPSC), organized by the Lunar and Planetary Institute in Houston. Held annually since 1970, occurs in March.

- Division for Planetary Sciences (DPS) meeting held annually since 1970 at a different location each year, predominantly within the mainland US. Occurs around October.

- American Geophysical Union (AGU) annual Fall meeting in December in San Francisco.

- American Geophysical Union (AGU) Joint Assembly (co-sponsored with other societies) in April–May, in various locations around the world.

- Meteoritical Society annual meeting, held during the Northern Hemisphere summer, generally alternating between North America and Europe.

- European Planetary Science Congress (EPSC), held annually around September at a location within Europe.

Smaller workshops and conferences on particular fields occur worldwide throughout the year.

Major Institutions

This non-exhaustive list includes those institutions and universities with major groups of people working in planetary science. Alphabetical order is used.

National Space Agencies

- Ames (NASA)

- Canadian Space Agency (CSA). Annual budget CAD $488.7 million (2013–2014).

- China National Space Administration (CNSA) (People's Republic of China). Budget $0.5-1.3 Billion (est.).

- Centre national d'études spatiales French National Centre of Space Research,Budget €1.920 Billion (2012).

- Deutsches Zentrum für Luft- und Raumfahrt e.V., (German: abbreviated DLR), the German Aerospace Center.Budget $2 Billion (2010).

- European Space Agency (ESA). Budget $5.51 Billion (2013).

- Russian Federal Space Agency Budget $5.61 Billion (2013).

- GSFC (NASA),

- Indian Space Research Organisation (ISRO),

- Israel Space Agency (ISA),

- Italian Space Agency Budget ~$1 Billion (2010).

- Japan Aerospace Exploration Agency (JAXA). Budget $2.15 Billion (2012).

- JPL (NASA),

- NASA: Considerable number of research groups, including the JPL, GSFC, Ames. Budget $18.724 Billion (2011).

- National Space Organization (Republic of China in Taiwan).

- UK Space Agency (UKSA).

Other Institutions

- The Australian National University's Planetary Science Institute

- Brown University Planetary Geosciences Group

- Caltech's Division of Geological and Planetary Sciences

- Cornell University's Space and Planetary Science

- Florida Institute of Technology's Department of Physics and Space Sciences

- Johns Hopkins University Applied Physics Laboratory
- Lunar and Planetary Institute
- MIT Dept. of Earth, Atmospheric and Planetary Sciences
- Open University Planetary and Space Sciences Research Institute
- Planetary Science Institute
- Stony Brook University's Geosciences Department and soon to open Center for Planetary Exploration
- UCL/Birkbeck's Centre for Planetary Sciences
- UCLA Dept. of Earth and Space Sciences
- University of Arizona's Lunar and Planetary Lab
- University of California Santa Cruz's Department of Earth & Planetary Sciences
- University of Hawaii's Hawaii Institute of Geophysics and Planetology
- University of Copenhagen's Center for Planetary Research
- University of Central Florida Planetary Sciences Group
- University of British Columbia Institute for Planetary Science
- University of Western Ontario's Centre for Planetary Science and Exploration
- University of Tennessee Department of Earth and Planetary Sciences
- University of Colorado's Department of Astrophysical and Planetary Sciences
- INAF– Istituto di Astrofisica e Planetologia Spaziali

Basic Concepts

- Asteroid
- Celestial mechanics
- Comet
- Dwarf planet
- Extrasolar planet
- Gas giant
- Icy moon
- Kuiper belt
- Magnetosphere
- Minor planet
- Planet
- Planetary differentiation
- Planetary system
- Definition of a planet
- Space weather
- Terrestrial planet

Earth Spheres: A Significant Topic

The planetary functions of the earth can be divided into many spheres. They have been divided for detailed scientific study. Some of these are the atmosphere, geosphere, biosphere etc. They provide vital energy to each other as well as enable life on earth. The major components of the Earth are discussed in this chapter.

Atmosphere of Earth

The atmosphere of Earth is the layer of gases, commonly known as air, that surrounds the planet Earth and is retained by Earth's gravity. The atmosphere protects life on Earth by absorbing ultraviolet solar radiation, warming the surface through heat retention (greenhouse effect), and reducing temperature extremes between day and night (the diurnal temperature variation).

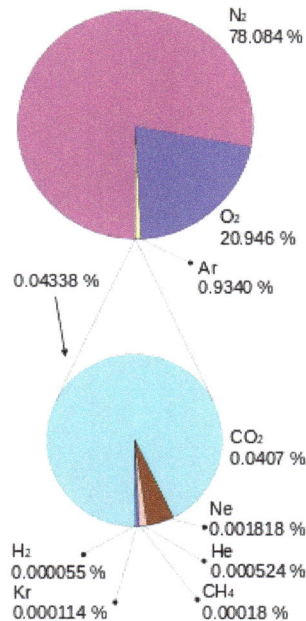

N_2
78.084 %

O_2
20.946 %

Ar
0.9340 %

0.04338 %

CO_2
0.0407 %

Ne
0.001818 %

He
0.000524 %

H_2
0.000055 %

Kr
0.000114 %

CH_4
0.00018 %

Composition of Earth's atmosphere by volume. The lower pie represents the trace gases that together compose about 0.038% of the atmosphere (0.043% with CO2 at 2014 concentration). The numbers are from a variety of years (mainly 1987, with CO2 and methane from 2009) and do not represent any single source.

By volume, dry air contains 78.09% nitrogen, 20.95% oxygen, 0.93% argon, 0.039% carbon dioxide, and small amounts of other gases. Air also contains a variable amount of water vapor, on average around 1% at sea level, and 0.4% over the entire atmosphere. Air content and atmospheric pressure vary at different layers, and air suitable for use in photosynthesis by terrestrial plants and breathing of terrestrial animals is found only in Earth's troposphere and in artificial atmospheres.

The atmosphere has a mass of about 5.15×1018 kg, three quarters of which is within about 11 km (6.8 mi; 36,000 ft) of the surface. The atmosphere becomes thinner and thinner with increasing altitude, with no definite boundary between the atmosphere and outer space. The Kármán line, at 100 km (62 mi), or 1.57% of Earth's radius, is often used as the border between the atmosphere and outer space. Atmospheric effects become noticeable during atmospheric reentry of spacecraft at an altitude of around 120 km (75 mi). Several layers can be distinguished in the atmosphere, based on characteristics such as temperature and composition.

Blue light is scattered more than other wavelengths by the gases in the atmosphere, giving Earth a blue halo when seen from space onboard *ISS at a height of* 402–424 km

The study of Earth's atmosphere and its processes is called atmospheric science (aerology). Early pioneers in the field include Léon Teisserenc de Bort and Richard Assmann.

Composition

Mean atmospheric water vapor

The three major constituents of air, and therefore of Earth's atmosphere, are nitrogen, oxygen, and argon. Water vapor accounts for roughly 0.25% of the atmosphere by mass. The concentration of water vapor (a greenhouse gas) varies significantly from around 10 ppm by volume in the coldest portions of the atmosphere to as much as 5% by volume in hot, humid air masses, and concentrations of other atmospheric gases are typically quoted in terms of dry air (without water vapor). The remaining gases are often referred to as trace gases, among which are the greenhouse gases, principally carbon dioxide, methane, nitrous oxide, and ozone. Filtered air includes trace amounts of many other chemical compounds. Many substances of natural origin may be present in locally and seasonally variable small amounts as aerosols in an unfiltered air sample, including dust of mineral and organic composition, pollen and spores, sea spray, and volcanic ash. Various industrial pollutants also may be present as gases or aerosols, such as chlorine (elemental or in compounds), fluorine compounds and elemental mercury vapor. Sulfur compounds such as hydrogen sulfide and sulfur dioxide (SO_2) may be derived from natural sources or from industrial air pollution.

Major constituents of dry air, by volume			
Gas		Volume[A]	
Name	Formula	in ppmv[B]	in %
Nitrogen	N_2	780,840	78.084
Oxygen	O_2	209,460	20.946
Argon	Ar	9,340	0.9340
Carbon dioxide	CO_2	397	0.0397
Neon	Ne	18.18	0.001818
Helium	He	5.24	0.000524
Methane	CH_4	1.79	0.000179
Not included in above dry atmosphere:			
Water vapor[C]	H_2O	10–50,000[D]	0.001%–5%[D]

notes:

[A] volume fraction is equal to mole fraction for ideal gas only, also see volume (thermodynamics)
[B] ppmv: parts per million by volume
[C] Water vapor is about 0.25% *by mass* over full atmosphere
[D] Water vapor strongly varies locally

Structure of the Atmosphere

Principal Layers

In general, air pressure and density decrease with altitude in the atmosphere. However, temperature has a more complicated profile with altitude, and may remain relatively constant or even increase with altitude in some regions. Because the general pattern of the temperature/altitude profile is constant and measurable by means of instrumented balloon soundings, the temperature behavior provides a useful metric to distinguish atmospheric layers. In this way, Earth's atmosphere can be divided (called atmospheric stratification) into five main layers. Excluding the exosphere, Earth has four primary layers, which are the troposphere, stratosphere, mesosphere, and thermosphere. From highest to lowest, the five main layers are:

- Exosphere: 700 to 10,000 km (440 to 6,200 miles)

- Thermosphere: 80 to 700 km (50 to 440 miles)

- Mesosphere: 50 to 80 km (31 to 50 miles)

- Stratosphere: 12 to 50 km (7 to 31 miles)

- Troposphere: 0 to 12 km (0 to 7 miles)

Earth's atmosphere Lower 4 layers of the atmosphere in 3 dimensions as seen diagonally from above the exobase. Layers drawn to scale, objects within the layers are not to scale. Aurorae shown here at the bottom of the thermosphere can actually form at any altitude in this atmospheric layer

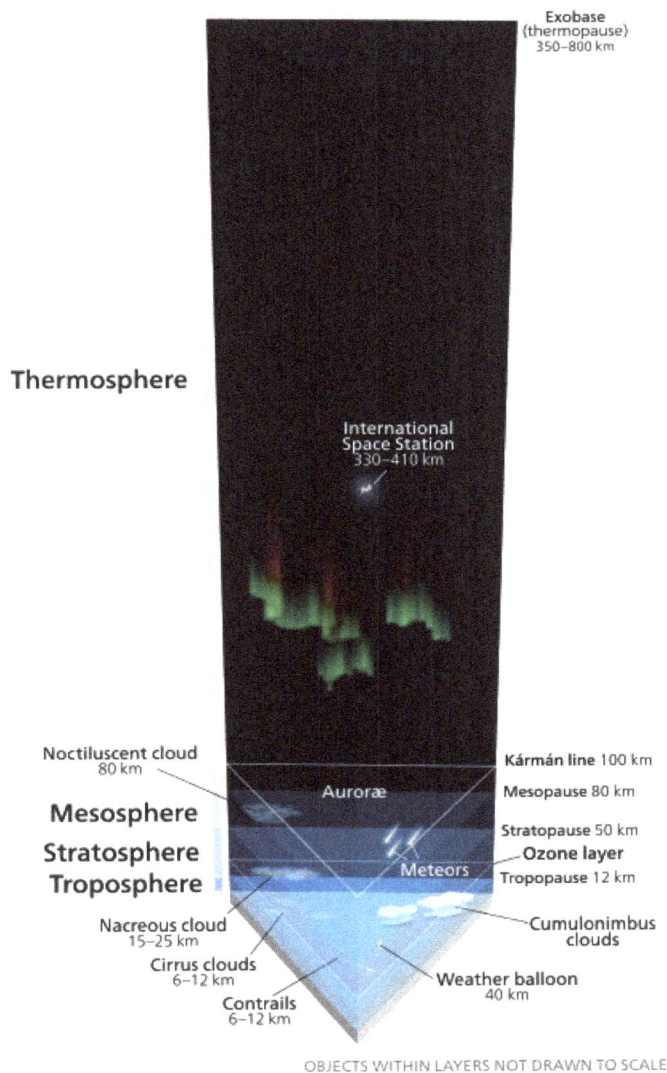

Exobase
(thermopause)
350–800 km

Thermosphere

International
Space Station
330–410 km

Noctiluscent cloud
80 km

Aurorae

Kármán line 100 km

Mesopause 80 km

Mesosphere

Stratosphere

Stratopause 50 km

Ozone layer

Troposphere

Meteors

Tropopause 12 km

Nacreous cloud
15–25 km

Cumulonimbus
clouds

Cirrus clouds
6–12 km

Weather balloon
40 km

Contrails
6–12 km

OBJECTS WITHIN LAYERS NOT DRAWN TO SCALE

Exosphere

The exosphere is the outermost layer of Earth's atmosphere (i.e. the upper limit of the atmosphere). It extends from the exobase, which is located at the top of the thermosphere at an altitude of about 700 km above sea level, to about 10,000 km (6,200 mi; 33,000,000 ft) where it merges into the solar wind.

This layer is mainly composed of extremely low densities of hydrogen, helium and several heavier molecules including nitrogen, oxygen and carbon dioxide closer to the exobase. The atoms and molecules are so far apart that they can travel hundreds of kilometers without colliding with one another. Thus, the exosphere no longer behaves like a gas, and the particles constantly escape into space. These free-moving particles follow ballistic trajectories and may migrate in and out of the magnetosphere or the solar wind.

The exosphere is located too far above Earth for any meteorological phenomena to be possible. However, the aurora borealis and aurora australis sometimes occur in the lower part of the exosphere, where they overlap into the thermosphere. The exosphere contains most of the satellites orbiting Earth.

Thermosphere

The thermosphere is the second-highest layer of Earth's atmosphere. It extends from the mesopause (which separates it from the mesosphere) at an altitude of about 80 km (50 mi; 260,000 ft) up to the thermopause at an altitude range of 500–1000 km (310–620 mi; 1,600,000–3,300,000 ft). The height of the thermopause varies considerably due to changes in solar activity. Because the thermopause lies at the lower boundary of the exosphere, it is also referred to as the exobase. The lower part of the thermosphere, from 80 to 550 kilometres (50 to 342 mi) above Earth's surface, contains the ionosphere.

The temperature of the thermosphere gradually increases with height. Unlike the stratosphere beneath it, wherein a temperature inversion is due to the absorption of radiation by ozone, the inversion in the thermosphere occurs due to the extremely low density of its molecules. The temperature of this layer can rise as high as 1500 °C (2700 °F), though the gas molecules are so far apart that its temperature in the usual sense is not very meaningful. The air is so rarefied that an individual molecule (of oxygen, for example) travels an average of 1 kilometre (0.62 mi; 3300 ft) between collisions with other molecules. Although the thermosphere has a high proportion of molecules with high energy, it would not feel hot to a human in direct contact, because its density is too low to conduct a significant amount of energy to or from the skin.

This layer is completely cloudless and free of water vapor. However non-hydrometeorological phenomena such as the aurora borealis and aurora australis are occasionally seen in the thermosphere. The International Space Station orbits in this layer, between 350 and 420 km (220 and 260 mi).

Mesosphere

The mesosphere is the third highest layer of Earth's atmosphere, occupying the region above the stratosphere and below the thermosphere. It extends from the stratopause at an altitude of about 50 km (31 mi; 160,000 ft) to the mesopause at 80–85 km (50–53 mi; 260,000–280,000 ft) above sea level.

Temperatures drop with increasing altitude to the mesopause that marks the top of this middle layer of the atmosphere. It is the coldest place on Earth and has an average temperature around −85 °C (−120 °F; 190 K).

Just below the mesopause, the air is so cold that even the very scarce water vapor at this altitude can be sublimated into polar-mesospheric noctilucent clouds. These are the highest clouds in the atmosphere and may be visible to the naked eye if sunlight reflects off them about an hour or two after sunset or a similar length of time before sunrise. They are most readily visible when the Sun is around 4 to 16 degrees below the horizon. A type of lightning referred to as either sprites or

ELVES, occasionally form far above tropospheric thunderclouds. The mesosphere is also the layer where most meteors burn up upon atmospheric entrance. It is too high above Earth to be accessible to jet-powered aircraft and balloons, and too low to permit orbital spacecraft. The mesosphere is mainly accessed by sounding rockets and rocket-powered aircraft.

Stratosphere

The stratosphere is the second-lowest layer of Earth's atmosphere. It lies above the troposphere and is separated from it by the tropopause. This layer extends from the top of the troposphere at roughly 12 km (7.5 mi; 39,000 ft) above Earth's surface to the stratopause at an altitude of about 50 to 55 km (31 to 34 mi; 164,000 to 180,000 ft).

The atmospheric pressure at the top of the stratosphere is roughly 1/1000 the pressure at sea level. It contains the ozone layer, which is the part of Earth's atmosphere that contains relatively high concentrations of that gas. The stratosphere defines a layer in which temperatures rise with increasing altitude. This rise in temperature is caused by the absorption of ultraviolet radiation (UV) radiation from the Sun by the ozone layer, which restricts turbulence and mixing. Although the temperature may be −60 °C (−76 °F; 210 K) at the tropopause, the top of the stratosphere is much warmer, and may be near 0 °C.

The stratospheric temperature profile creates very stable atmospheric conditions, so the stratosphere lacks the weather-producing air turbulence that is so prevalent in the troposphere. Consequently, the stratosphere is almost completely free of clouds and other forms of weather. However, polar stratospheric or nacreous clouds are occasionally seen in the lower part of this layer of the atmosphere where the air is coldest. This is the highest layer that can be accessed by jet-powered aircraft.

Troposphere

The troposphere is the lowest layer of Earth's atmosphere. It extends from Earth's surface to an average height of about 12 km, although this altitude actually varies from about 9 km (30,000 ft) at the poles to 17 km (56,000 ft) at the equator, with some variation due to weather. The troposphere is bounded above by the tropopause, a boundary marked in most places by a temperature inversion (i.e. a layer of relatively warm air above a colder one), and in others by a zone which is isothermal with height.

Although variations do occur, the temperature usually declines with increasing altitude in the troposphere because the troposphere is mostly heated through energy transfer from the surface. Thus, the lowest part of the troposphere (i.e. Earth's surface) is typically the warmest section of the troposphere. This promotes vertical mixing. *The troposphere contains roughly 80% of the* mass of Earth's atmosphere. The troposphere is denser than all its overlying atmospheric layers because a larger atmospheric weight sits on top of the troposphere and causes it to be most severely compressed. Fifty percent of the total mass of the atmosphere is located in the lower 5.6 km (18,000 ft) of the troposphere.

Nearly all atmospheric water vapor or moisture is found in the troposphere, so it is the layer where most of Earth's weather takes place. It has basically all the weather-associated cloud genus types generated by active wind circulation, although very tall cumulonimbus thunder clouds can penetrate the tropopause from below and rise into the lower part of the stratosphere. Most conventional aviation activity takes place in the troposphere, and it is the only layer that can be accessed by propeller-driven aircraft.

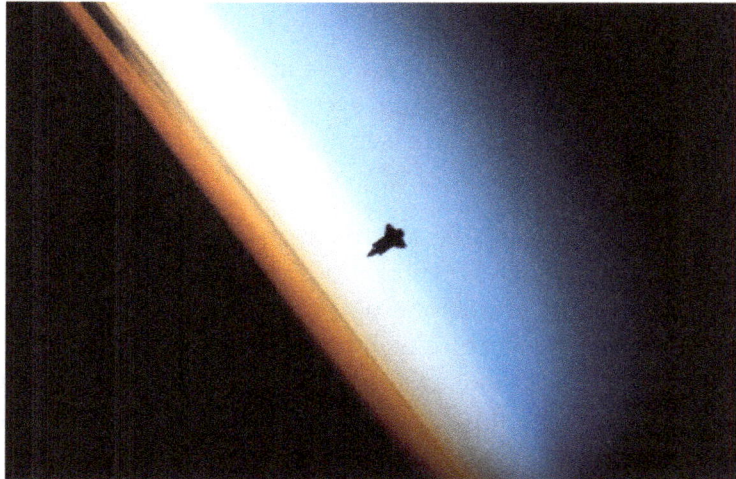

Space Shuttle Endeavour orbiting in the thermosphere. Because of the angle of the photo, it appears to straddle the stratosphere and mesosphere that actually lie more than 250 km below. The orange layer is the troposphere, which gives way to the whitish stratosphere and then the blue mesosphere.

Other Layers

Within the five principal layers that are largely determined by temperature, several secondary layers may be distinguished by other properties:

- The ozone layer is contained within the stratosphere. In this layer ozone concentrations are about 2 to 8 parts per million, which is much higher than in the lower atmosphere but still very small compared to the main components of the atmosphere. It is mainly located in the lower portion of the stratosphere from about 15–35 km (9.3–21.7 mi; 49,000–115,000 ft), though the thickness varies seasonally and geographically. About 90% of the ozone in Earth's atmosphere is contained in the stratosphere.

- The ionosphere is a region of the atmosphere that is ionized by solar radiation. It is responsible for auroras. During daytime hours, it stretches from 50 to 1,000 km (31 to 621 mi; 160,000 to 3,280,000 ft) and includes the mesosphere, thermosphere, and parts of the exosphere. However, ionization in the mesosphere largely ceases during the night, so auroras are normally seen only in the thermosphere and lower exosphere. The ionosphere forms the inner edge of the magnetosphere. It has practical importance because it influences, for example, radio propagation on Earth.

- The homosphere and heterosphere are defined by whether the atmospheric gases are well mixed. The surface-based homosphere includes the troposphere, stratosphere, mesosphere, and the lowest part of the thermosphere, where the chemical composition of the atmosphere does not depend on molecular weight because the gases are mixed by turbulence. This relatively homogeneous layer ends at the *turbopause found at about 100 km (62 mi; 330,000 ft), which places it about 20 km (12 mi; 66,000 ft) above the mesopause.*

Above this altitude lies the heterosphere, which includes the exosphere and most of the thermosphere. Here, the chemical composition varies with altitude. This is because the distance that particles can move without colliding with one another is large compared with the size of motions that cause mixing. This allows the gases to stratify by molecular weight, with the heavier ones, such as oxygen and nitrogen, present only near the bottom of the heterosphere. The upper part of the heterosphere is composed almost completely of hydrogen, the lightest element.[*clarification needed*]

- The planetary boundary layer is the part of the troposphere that is closest to Earth's surface and is directly affected by it, mainly through turbulent diffusion. During the day the planetary boundary layer usually is well-mixed, whereas at night it becomes stably stratified with weak or intermittent mixing. The depth of the planetary boundary layer ranges from as little as about 100 meters on clear, calm nights to 3000 m or more during the afternoon in dry regions.

The average temperature of the atmosphere at Earth's surface is 14 °C (57 °F; 287 K) or 15 °C (59 °F; 288 K), depending on the reference.

Physical Properties

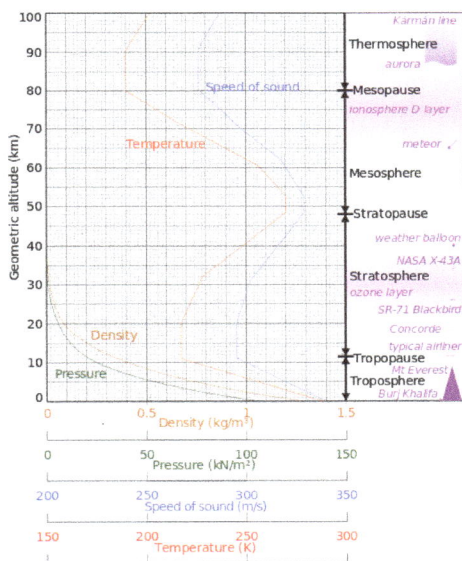

Comparison of the 1962 US Standard Atmosphere graph of geometric altitude against air density, pressure, the speed of sound and temperature with approximate altitudes of various objects.

Pressure and Thickness

The average atmospheric pressure at sea level is defined by the International Standard Atmosphere as 101325 pascals (760.00 Torr; 14.6959 psi; 760.00 mmHg). This is sometimes referred to as a unit of standard atmospheres (atm). Total atmospheric mass is 5.1480×10^{18} kg (1.135×10^{19} lb), about 2.5% less than would be inferred from the average sea level pressure and Earth's area of

51007.2 megahectares, this portion being displaced by Earth's mountainous terrain. Atmospheric pressure is the total weight of the air above unit area at the point where the pressure is measured. Thus air pressure varies with location and weather.

If the entire mass of the atmosphere had a uniform density from sea level, it would terminate abruptly at an altitude of 8.50 km (27,900 ft). It actually decreases exponentially with altitude, dropping by half every 5.6 km (18,000 ft) or by a factor of 1/e every 7.64 km (25,100 ft), the average scale height of the atmosphere below 70 km (43 mi; 230,000 ft). However, the atmosphere is more accurately modeled with a customized equation for each layer that takes gradients of temperature, molecular composition, solar radiation and gravity into account.

In summary, the mass of Earth's atmosphere is distributed approximately as follows:

- 50% is below 5.6 km (18,000 ft).

- 90% is below 16 km (52,000 ft).

- 99.99997% is below 100 km (62 mi; 330,000 ft), the Kármán line. By international convention, this marks the beginning of space where human travelers are considered astronauts.

By comparison, the summit of Mt. Everest is at 8,848 m (29,029 ft); commercial airliners typically cruise between 10 km (33,000 ft) and 13 km (43,000 ft) where the thinner air improves fuel economy; weather balloons reach 30.4 km (100,000 ft) and above; and the highest X-15 flight in 1963 reached 108.0 km (354,300 ft).

Even above the Kármán line, significant atmospheric effects such as auroras still occur. Meteors begin to glow in this region, though the larger ones may not burn up until they penetrate more deeply. The various layers of Earth's ionosphere, important to HF radio propagation, begin below 100 km and extend beyond 500 km. By comparison, the International Space Station and Space Shuttle typically orbit at 350–400 km, within the F-layer of the ionosphere where they encounter enough atmospheric drag to require reboosts every few months. Depending on solar activity, satellites can experience noticeable atmospheric drag at altitudes as high as 700–800 km.

Temperature and Speed of Sound

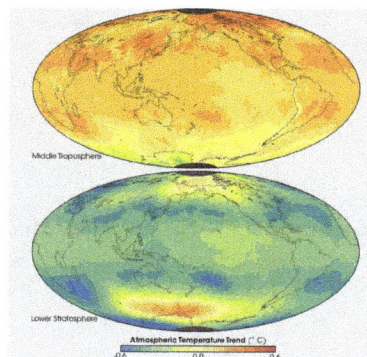

These images show temperature trends in two thick layers of the atmosphere as measured by a series of satellite-based instruments between January 1979 and December 2005. The measurements were taken by Microwave Sounding Units and Advanced Microwave Sounding Units flying on a series of National Oceanic and Atmospheric Administration (NOAA) weather satellites. The instruments record microwaves emitted from oxygen molecules in the atmosphere. Source:

The division of the atmosphere into layers mostly by reference to temperature is discussed above. Temperature decreases with altitude starting at sea level, but variations in this trend begin above 11 km, where the temperature stabilizes through a large vertical distance through the rest of the troposphere. In the stratosphere, starting above about 20 km, the temperature increases with height, due to heating within the ozone layer caused by capture of significant ultraviolet radiation from the Sun by the dioxygen and ozone gas in this region. Still another region of increasing temperature with altitude occurs at very high altitudes, in the aptly-named thermosphere above 90 km.

Because in an ideal gas of constant composition the speed of sound depends only on temperature and not on the gas pressure or density, the speed of sound in the atmosphere with altitude takes on the form of the complicated temperature profile and does not mirror altitudinal changes in density or pressure.

Density and Mass

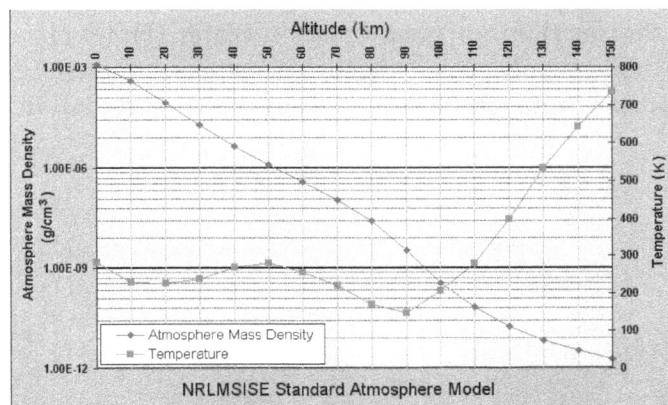

Temperature and mass density against altitude from the NRLMSISE-00 standard atmosphere model (the eight dotted lines in each "decade" are at the eight cubes 8, 27, 64, …, 729)

The density of air at sea level is about 1.2 kg/m3 (1.2 g/L, 0.0012 g/cm3). Density is not measured directly but is calculated from measurements of temperature, pressure and humidity using the equation of state for air (a form of the ideal gas law). Atmospheric density decreases as the altitude increases. This variation can be approximately modeled using the barometric formula. More sophisticated models are used to predict orbital decay of satellites.

The average mass of the atmosphere is about 5 quadrillion (5×10^{15}) tonnes or 1/1,200,000 the mass of Earth. According to the American National Center for Atmospheric Research, "The total mean mass of the atmosphere is 5.1480×10^{18} kg with an annual range due to water vapor of 1.2 or 1.5×10^{15} kg, depending on whether surface pressure or water vapor data are used; somewhat smaller than the previous estimate. The mean mass of water vapor is estimated as 1.27×10^{16} kg and the dry air mass as $5.1352 \pm 0.0003 \times 10^{18}$ kg."

Optical Properties

Solar radiation (or sunlight) is the energy Earth receives from the Sun. Earth also emits radiation back into space, but at longer wavelengths that we cannot see. Part of the incoming and emitted radiation is absorbed or reflected by the atmosphere.

Scattering

When light passes through Earth's atmosphere, photons interact with it through *scattering*. If the light does not interact with the atmosphere, it is called *direct radiation and* is what you see if you were to look directly at the Sun. *Indirect radiation is light that has been scattered in the atmosphere. For example, on an* overcast day when you cannot see your shadow there is no direct radiation reaching you, it has all been scattered. As another example, due to a phenomenon called Rayleigh scattering, shorter (blue) wavelengths scatter more easily than longer (red) wavelengths. This is why the sky looks blue; you are seeing scattered blue light. This is also why sunsets are red. Because the Sun is close to the horizon, the Sun's rays pass through more atmosphere than normal to reach your eye. Much of the blue light has been scattered out, leaving the red light in a sunset.

Absorption

Different molecules absorb different wavelengths of radiation. For example, O_2 and O_3 absorb almost all wavelengths shorter than 300 nanometers. Water (H_2O) absorbs many wavelengths above 700 nm. When a molecule absorbs a photon, it increases the energy of the molecule. This heats the atmosphere, but the atmosphere also cools by emitting radiation, as discussed below.

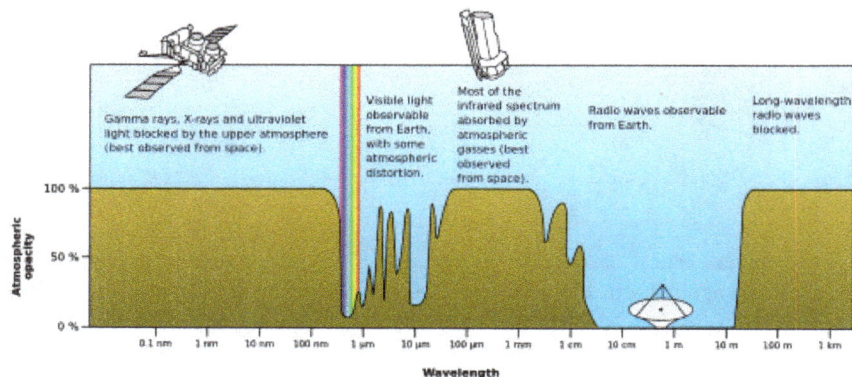

Rough plot of Earth's atmospheric transmittance (or opacity) to various wavelengths of electromagnetic radiation, including visible light.

The combined absorption spectra of the gases in the atmosphere leave "windows" of low opacity, allowing the transmission of only certain bands of light. The optical window runs from around 300 nm (ultraviolet-C) up into the range humans can see, the visible spectrum (commonly called light), at roughly 400–700 nm and continues to the infrared to around 1100 nm. There are also infrared and radio windows that transmit some infrared and radio waves at longer wavelengths. For example, the radio window runs from about one centimeter to about eleven-meter waves.

Emission

Emission is the opposite of absorption, it is when an object emits radiation. Objects tend to emit amounts and wavelengths of radiation depending on their "black body" emission curves, therefore hotter objects tend to emit more radiation, with shorter wavelengths. Colder objects emit less radiation, with longer wavelengths. For example, the Sun is approximately 6,000 K (5,730 °C; 10,340 °F), its radiation peaks near 500 nm, and is visible to the human eye. Earth is approximately 290 K (17 °C; 62 °F), so its radiation peaks near 10,000 nm, and is much too long to be visible to humans.

Because of its temperature, the atmosphere emits infrared radiation. For example, on clear nights Earth's surface cools down faster than on cloudy nights. This is because clouds (H_2O) are strong absorbers and emitters of infrared radiation. This is also why it becomes colder at night at higher elevations.

The greenhouse effect is directly related to this absorption and emission effect. Some gases in the atmosphere absorb and emit infrared radiation, but do not interact with sunlight in the visible spectrum. Common examples of these are CO_2 and H_2O.

Refractive Index

The refractive index of air is close to, but just greater than 1. Systematic variations in refractive index can lead to the bending of light rays over long optical paths. One example is that, under some circumstances, observers onboard ships can see other vessels just over the horizon because light is refracted in the same direction as the curvature of Earth's surface.

The refractive index of air depends on temperature, giving rise to refraction effects when the temperature gradient is large. An example of such effects is the mirage.

Circulation

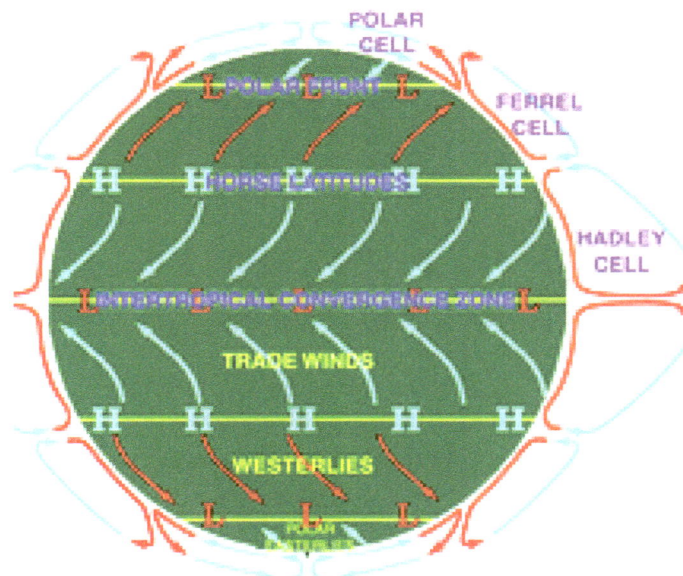

An idealised view of three large circulation cells.

Atmospheric circulation is the large-scale movement of air through the troposphere, and the means (with ocean circulation) by which heat is distributed around Earth. The large-scale structure of the atmospheric circulation varies from year to year, but the basic structure remains fairly constant because it is determined by Earth's rotation rate and the difference in solar radiation between the equator and poles.

Evolution of Earth's Atmosphere

Earliest Atmosphere

The first atmosphere would have consisted of gases in the solar nebula, primarily hydrogen. In addition, there would probably have been simple hydrides such as those now found in the gas giants (Jupiter and Saturn), notably water vapor, methane and ammonia. As the solar nebula dissipated, these gases would have escaped, partly driven off by the solar wind.

Second Atmosphere

Outgassing from volcanism, supplemented by gases produced during the late heavy bombardment of Earth by huge asteroids, produced the next atmosphere, consisting largely of nitrogen plus carbon dioxide and inert gases. A major part of carbon-dioxide emissions soon dissolved in water and built up carbonate sediments.[*clarification needed*]

Researchers have found water-related sediments dating from as early as 3.8 billion years ago. About 3.4 billion years ago, nitrogen formed the major part of the then stable "second atmosphere". An influence of life has to be taken into account rather soon in the history of the atmosphere, because hints of early life-forms appear as early as 3.5 billion years ago. How Earth at that time maintained a climate warm enough for liquid water and life, if the early Sun put out 30% lower solar radiance than today, is a puzzle known as the "faint young Sun paradox".

The geological record however shows a continually relatively warm surface during the complete early temperature record of Earth - with the exception of one cold glacial phase about 2.4 billion years ago. In the late Archean eon an oxygen-containing atmosphere began to develop, apparently produced by photosynthesizing cyanobacteria which have been found as stromatolite fossils from 2.7 billion years ago. The early basic carbon isotopy (isotope ratio proportions) very much approximates current conditions, suggesting that the fundamental features of the carbon cycle became established as early as 4 billion years ago.

Ancient sediments in the Gabon dating from between about 2,150 and 2,080 million years ago provide a record of Earth's dynamic oxygenation evolution. These fluctuations in oxygenation were likely driven by the Lomagundi carbon isotope excursion.

Third Atmosphere

The constant re-arrangement of continents by plate tectonics influences the long-term evolution of the atmosphere by transferring carbon dioxide to and from large continental carbonate stores. Free oxygen did not exist in the atmosphere until about 2.4 billion years ago during the Great Oxygenation Event and its appearance is indicated by the end of the banded iron formations. Before this time, any oxygen produced by photosynthesis was consumed by oxidation of reduced materials, notably iron. Molecules of free oxygen did not start to accumulate in the atmosphere until the rate of production of oxygen began to exceed the availability of reducing materials. This point signifies a shift from a reducing atmosphere to an oxidizing atmosphere. O2 showed major variations until reaching a steady state of more than 15% by the end of the Precambrian. The following time span from 541 million years ago to the present day is the Phanerozoic eon, during the earliest period of which, the Cambrian, oxygen-requiring metazoan life forms began to appear.

Oxygen Content of Earth's Atmosphere

During the Course of the Last Billion Years

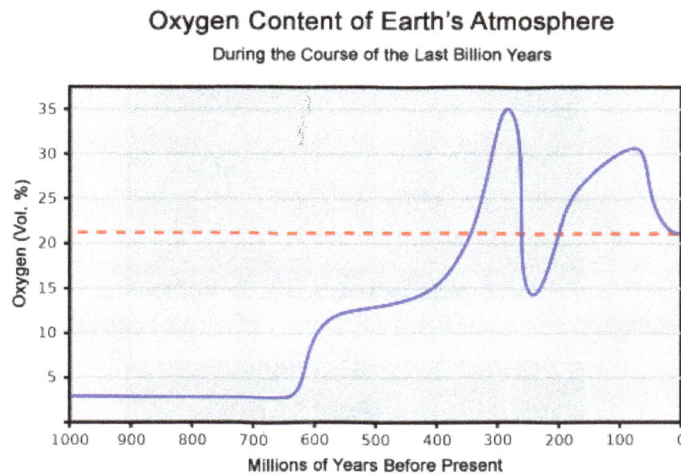

Oxygen content of the atmosphere over the last billion years. This diagram in more detail

The amount of oxygen in the atmosphere has fluctuated over the last 600 million years, reaching a peak of about 30% around 280 million years ago, significantly higher than today's 21%. Two main processes govern changes in the atmosphere: Plants use carbon dioxide from the atmosphere, releasing oxygen. Breakdown of pyrite and volcanic eruptions release sulfur into the atmosphere, which oxidizes and hence reduces the amount of oxygen in the atmosphere. However, volcanic eruptions also release carbon dioxide, which plants can convert to oxygen. The exact cause of the variation of the amount of oxygen in the atmosphere is not known. Periods with much oxygen in the atmosphere are associated with rapid development of animals. Today's atmosphere contains 21% oxygen, which is high enough for this rapid development of animals.

The scientific consensus is that the anthropogenic greenhouse gases currently accumulating in the atmosphere are the main cause of global warming.

Air Pollution

Air pollution is the introduction into the atmosphere of chemicals, particulate matter or biological materials that cause harm or discomfort to organisms. Stratospheric ozone depletion is caused by air pollution, chiefly from chlorofluorocarbons and other ozone-depleting substances.

Images from Space

Blue light is scattered more than other wavelengths by the gases in the atmosphere,

giving Earth a blue halo when seen from space.

The geomagnetic storms cause beautiful displays of aurora across the atmosphere.

Limb view, of Earth's atmosphere. Colors roughly denote the layers of the atmosphere.

This image shows the Moon at the centre, with the limb of Earth near the bottom transitioning into the orange-colored troposphere. The troposphere ends abruptly at the tropopause, which appears in the image as the sharp boundary between the orange- and blue-colored atmosphere. The silvery-blue noctilucent clouds extend far above Earth's troposphere.

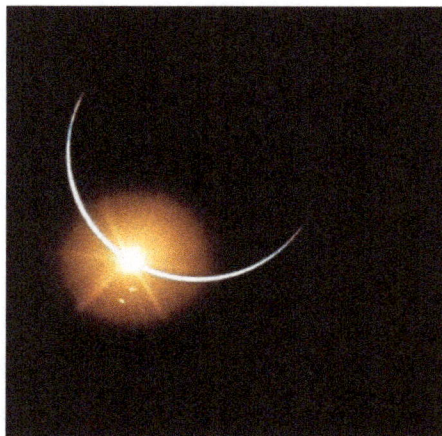

Earth's atmosphere backlit by the Sun in an eclipse observed from deep space onboard Apollo 12 in 1969.

On October 19, 2015 NASA started a website containing daily images of the full sunlit side of Earth on http://epic.gsfc.nasa.gov/. The images are taken from the Deep Space Climate Observatory (DSCOVR) and show Earth as it rotates during a day.

Hydrosphere

It has been estimated that there are 1386 million cubic kilometres of water on Earth. This includes water in liquid and frozen forms in groundwater, oceans, lakes and streams. Saltwater accounts for 97.5% of this amount. Fresh water accounts for only 2.5%. Of this fresh water, 68.7% is in the form of ice and permanent snow cover in the Arctic, the Antarctic, and mountain glaciers. 29.9% is in the form of fresh groundwater. Only 0.26% of the fresh water on Earth is in easily accessible lakes, reservoirs and river systems. The total mass of the Earth's hydrosphere is about 1.4×10^{18} tonnes, which is about 0.023% of Earth's total mass. About 20×10^{12} tonnes of this is in Earth's atmosphere (for practical purposes, 1 cubic metre of water weighs one tonne). Approximately 75% of Earth's surface, an area of some 361 million square kilometers (139.5 million square miles), is covered by ocean. The average salinity of Earth's oceans is about 35 grams of salt per kilogram of sea water (3.5%).

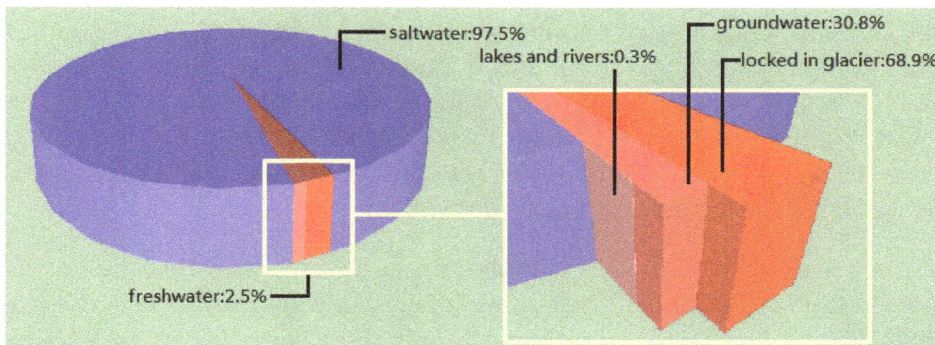

The hydrosphere is the combined mass of water found on, under, and above the surface of a planet.

Water Cycle

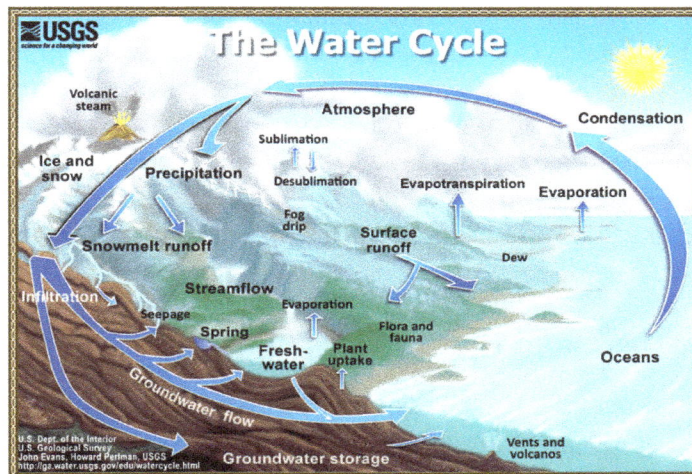

The hydrological cycle transfers water from one state or reservoir to another. Reservoirs include atmospheric moisture (snow, rain and clouds),streams, oceans, rivers, lakes, groundwater, sub-terranean aquifers, polar icecaps and saturated soil. Solar energy, in the form of heat and light (insolation), and gravity cause the transfer from one state to another over periods from hours to thousands of years. Most evaporation comes from the oceans and is returned to the earth as snow or rain (page 27). Sublimation refers to evaporation from snow and ice. Transpiration refers to the expiration of water through the minute pores or stomata of trees. Evapotranspiration is the term used by hydrologists in reference to the three processes together, transpiration, sublimation and evaporation.

In his book *Water,* Marq de Villiers described the hydrosphere as a closed system in which water exists. The hydrosphere is intricate, complex, interdependent, all-pervading and stable and "seems purpose-built for regulating life (de Villiers 2003:26)." De Villiers claimed that, "On earth, the total amount of water has almost certainly not changed since geological times: what we had then we still have. Water can be polluted, abused, and misused but it is neither created nor destroyed, it only migrates. There is no evidence that water vapor escapes into space (page 26)."

"Every year the turnover of water on Earth involves 577,000 km3 of water. This is water that evaporates from the oceanic surface (502,800 km3) and from land (74,200 km3). The same amount of water falls as atmospheric precipitation, 458,000 km3 on the ocean and 119,000 km3 on land. The difference between precipitation and evaporation from the land surface (119,000 - 74,200 = 44,800 km3/year) represents the total runoff of the Earth's rivers (42,700 km3/year) and direct groundwater runoff to the ocean (2100 km3/year). These are the principal sources of fresh water to support life necessities and man's economic activities."

Water is a basic necessity of life. Since 2/3 of the Earth is covered by water, the Earth is also called the blue planet and the watery planet.Hydrosphere plays an important role in the existence of the atmosphere in its present form. Oceans are important in this regard. When the Earth was formed it had only a very thin atmosphere rich in hydrogen and helium similar to the present atmosphere of Mercury. Later the gases hydrogen and helium were expelled from the atmosphere. The gases and water vapor released as the Earth cooled became our present atmosphere. Other gases and water vapor released by volcanoes also entered the atmosphere. As the Earth cooled the water vapor in the atmosphere condensed and fell as rain. The atmosphere cooled further as atmospheric carbon dioxide dissolved in to rain water. In turn this further caused the water vapor to condense and fall as rain. This rain water filled the depressions on the Earth's surface and formed the oceans. It is estimated that this occurred about 4000 million years ago. The first life forms began in the oceans. These organisms did not breathe oxygen. Later, when cyanobacteria evolved, the process of conversion of carbon dioxide into food and oxygen began. As a result, our atmosphere has a distinctly different composition from that of the other planets; it is a fundamental requirement for life on Earth.

Recharging Reservoirs

According to Igor A. Shiklomanov, it takes 2500 years for the complete recharge and replenishment of oceanic waters, 10,000 years for permafrost and ice, 1500 years for deep groundwater and mountainous glaciers, 17 years in lakes and 16 days in rivers.

Specific Fresh Water Availability

"Specific water availability is the residual (after use) per capita quantity of fresh water." Fresh water resources are unevenly distributed in terms of space and time and can go from floods to water shortages within months in the same area. In 1998 76% of the total population had a specific water availability of less than 5.0 thousand m3 per year per capita. Already by 1998, 35% of the global population suffered "very low or catastrophically low water supplies" and Shiklomanov predicted that the situation would deteriorate in the twenty-first century with "most of the Earth's population will be living under the conditions of low or catastrophically low water supply" by 2025. There is only 2.5% of fresh water in the hydrosphere.

Lithosphere

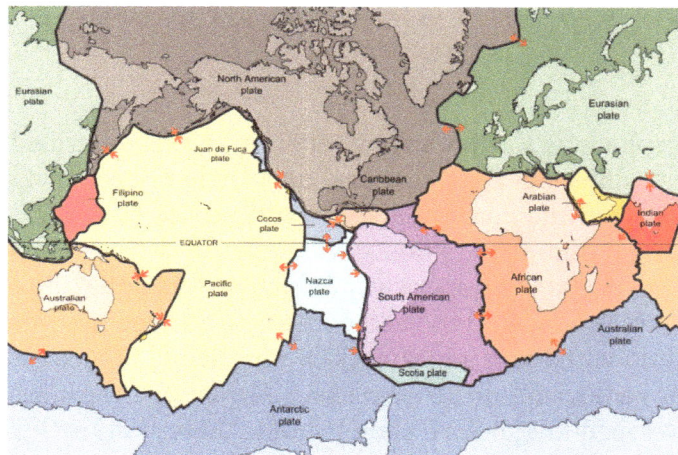

The tectonic plates of the lithosphere on Earth

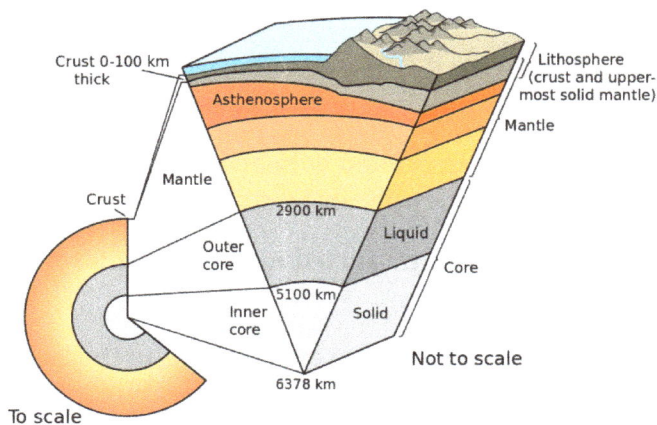

Earth cutaway from core to crust, the lithosphere comprising the crust and lithospheric mantle (detail not to scale)

A lithosphere is the rigid, outermost shell of a terrestrial-type planet or natural satellite that is defined by its rigid mechanical properties. On Earth, it is composed of the crust and the portion of the upper mantle that behaves elastically on time scales of thousands of years or greater. The outermost shell of a rocky planet, the crust, is defined on the basis of its chemistry and mineralogy.

Earth's Lithosphere

Earth's lithosphere includes the crust and the uppermost mantle, which constitute the hard and rigid outer layer of the Earth. The lithosphere is subdivided into tectonic plates. The uppermost part of the lithosphere that chemically reacts to the atmosphere, hydrosphere and biosphere through the soil forming process is called the pedosphere. The lithosphere is underlain by the asthenosphere which is the weaker, hotter, and deeper part of the upper mantle. The boundary between the lithosphere and the underlying asthenosphere is known as the Lithosphere-Asthenosphere boundary and is defined by a difference in response to stress: the lithosphere remains rigid for very long periods of geologic time in which it deforms elastically and through brittle failure, while the asthenosphere deforms viscously and accommodates strain through plastic deformation. The study of past and current formations of landscapes is called geomorphology.

History

The concept of the lithosphere as Earth's strong outer layer was described by A.E.H. Love in his 1911 monograph "Some problems of Geodynamics" and further developed by Joseph Barrell, who wrote a series of papers about the concept and introduced the term "lithosphere". The concept was based on the presence of significant gravity anomalies over continental crust, from which he inferred that there must exist a strong upper layer (which he called the lithosphere) above a weaker layer which could flow (which he called the asthenosphere). These ideas were expanded by Reginald Aldworth Daly in 1940 with his seminal work "Strength and Structure of the Earth" and have been broadly accepted by geologists and geophysicists. Although these ideas about lithosphere and asthenosphere were developed long before plate tectonic theory was articulated in the 1960s, the concepts that a strong lithosphere exists and that this rests on a weak asthenosphere are essential to that theory.

Types

There are two types of lithosphere:

- Oceanic lithosphere, which is associated with oceanic crust and exists in the ocean basins (mean density of about 2.9 grams per cubic centimeter)

- Continental lithosphere, which is associated with continental crust (mean density of about 2.7 grams per cubic centimeter)

The thickness of the lithosphere is considered to be the depth to the isotherm associated with the transition between brittle and viscous behavior. The temperature at which olivine begins to deform viscously (~1000 °C) is often used to set this isotherm because olivine is generally the weakest mineral in the upper mantle. Oceanic lithosphere is typically about 50–140 km thick (but beneath the mid-ocean ridges is no thicker than the crust), while continental lithosphere has a range in thickness from about 40 km to perhaps 280 km; the upper ~30 to ~50 km of typical continental lithosphere is crust. The mantle part of the lithosphere consists largely of peridotite. The crust is distinguished from the upper mantle by the change in chemical composition that takes place at the Moho discontinuity.

Oceanic Lithosphere

Oceanic lithosphere consists mainly of mafic crust and ultramafic mantle (peridotite) and is denser than continental lithosphere, for which the mantle is associated with crust made of felsic rocks. Oceanic lithosphere thickens as it ages and moves away from the mid-ocean ridge. This thickening occurs by conductive cooling, which converts hot asthenosphere into lithospheric mantle and causes the oceanic lithosphere to become increasingly thick and dense with age. The thickness of the mantle part of the oceanic lithosphere can be approximated as a thermal boundary layer that thickens as the square root of time.

$$h \sim 2\sqrt{\kappa t}$$

Here, h is the thickness of the oceanic mantle lithosphere, κ is the thermal diffusivity (approximately $10-6$ m2/s) for silicate rocks, and t is the age of the given part of the lithosphere. The age is often equal to L/V, where L is the distance from the spreading centre of mid-oceanic ridge, and V is velocity of the lithospheric plate.

Oceanic lithosphere is less dense than asthenosphere for a few tens of millions of years but after this becomes increasingly denser than asthenosphere. This is because the chemically differentiated oceanic crust is lighter than asthenosphere, but thermal contraction of the mantle lithosphere makes it more dense than the asthenosphere. The gravitational instability of mature oceanic lithosphere has the effect that at subduction zones, oceanic lithosphere invariably sinks underneath the overriding lithosphere, which can be oceanic or continental. New oceanic lithosphere is constantly being produced at mid-ocean ridges and is recycled back to the mantle at subduction zones. As a result, oceanic lithosphere is much younger than continental lithosphere: the oldest oceanic lithosphere is about 170 million years old, while parts of the continental lithosphere are billions of years old. The oldest parts of continental lithosphere underlie cratons, and the mantle lithosphere there is thicker and less dense than typical; the relatively low density of such mantle "roots of cratons" helps to stabilize these regions.

Subducted Lithosphere

Geophysical studies in the early 21st century posit that large pieces of the lithosphere have been subducted into the mantle as deep as 2900 km to near the core-mantle boundary, while others "float" in the upper mantle, while some stick down into the mantle as far as 400 km but remain "attached" to the continental plate above, similar to the extent of the "tectosphere" proposed by Jordan in 1988.

Mantle Xenoliths

Geoscientists can directly study the nature of the subcontinental mantle by examining mantle xenoliths brought up in kimberlite, lamproite, and other volcanic pipes. The histories of these xenoliths have been investigated by many methods, including analyses of abundances of isotopes of osmium and rhenium. Such studies have confirmed that mantle lithospheres below some cratons have persisted for periods in excess of 3 billion years, despite the mantle flow that accompanies plate tectonics.

Pedosphere

The pedosphere is the outermost layer of the Earth that is composed of soil and subject to soil formation processes. It exists at the interface of the lithosphere, atmosphere, hydrosphere and biosphere. The sum total of all the organisms, soils, water and air is termed as the "pedosphere". The pedosphere is the skin of the Earth and only develops when there is a dynamic interaction between the atmosphere (air in and above the soil), biosphere (living organisms), lithosphere (unconsolidated regolith and con-solidated bedrock) and the hydrosphere (water in, on and below the soil). The pedosphere is the foundation of terrestrial life on this planet. There is a realization that the pedosphere needs to be distinctly recognized as a dynamic interface of all terrestrial ecosystems and be integrated into the Earth System Science knowledge base.

The pedosphere acts as the mediator of chemical and biogeochemical flux into and out of these respective systems and is made up of gaseous, mineralic, fluid and biologic components. The pedosphere lies within the Critical Zone, a broader interface that includes vegetation, pedosphere, groundwater aquifer systems, regolith and finally ends at some depth in the bedrock where the biosphere and hydrosphere cease to make significant changes to the chemistry at depth. As part of the larger global system, any particular environment in which soil forms is influenced solely by its geographic position on the globe as climatic, geologic, biologic and anthropogenic changes occur with changes in longitude and latitude.

The pedosphere lies below the vegetative cover of the biosphere and above the hydrosphere and lithosphere. The soil forming process (pedogenesis) can begin without the aid of biology but is significantly quickened in the presence of biologic reactions. Soil formation begins with the chemical and/or physical breakdown of minerals to form the initial material that overlies the bedrock substrate. Biology quickens this by secreting acidic compounds (dominantly fulvic acids) that help break rock apart. Particular biologic pioneers are lichen, mosses and seed bearing plants but many other inorganic reactions take place that diversify the chemical makeup of the early soil layer. Once weathering and decomposition products accumulate, a coherent soil body allows the migration of fluids both vertically and laterally through the soil profile causing ion exchange between solid, fluid and gaseous phases. As time progresses, the bulk geochemistry of the soil layer will deviate away from the initial composition of the bedrock and will evolve to a chemistry that reflects the type of reactions that take place in the soil.

Lithosphere

The primary conditions for soil development are controlled by the chemical composition of the rock that the soil will eventually be forming on. Rock types that form the base of the soil profile are often either sedimentary (carbonate or siliceous), igneous or metaigneous (metamorphosed igneous rocks) or volcanic and metavolcanic rocks. The rock type and the processes that lead to its exposure at the surface are controlled by the regional geologic setting of the specific area under study, which revolve around the underlying theory of plate tectonics, subsequent deformation, uplift, subsidence and deposition.

Metaigneous and metavolcanic rocks form the largest component of cratons and are high in silica. Igneous and volcanic rocks are also high in silica but with non-metamorphosed rock, weathering becomes faster and the mobilization of ions is more widespread. Rocks high in silica produce silicic acid as a weathering product. There are few rock types that lead to localized enrichment of some of the biologically limiting elements like phosphorus (P) and nitrogen (N). Phosphatic shale (<15% P_2O_5) and phosphorite (>15% P_2O_5) form in anoxic deep water basins that preserve organic material. Greenstone (metabasalt), phyllite and schist release up to 30-50% of the nitrogen pool. Thick successions of carbonate rocks are often deposited on craton margins during sea level rise. The widespread dissolution of carbonate and evaporate minerals leads to elevated levels of $Mg2+$, $HCO3-$, $Sr2+$, $Na+$, $Cl-$ and $SO42-$ ions in aqueous solution.

Weathering and Dissolution of Minerals

The process of soil formation is dominated by chemical weathering of silicate minerals, aided by acidic products of pioneering plants and organisms as well as carbonic acid inputs from the atmosphere. Carbonic acid is produced in the atmosphere and soil layers through the carbonation reaction.

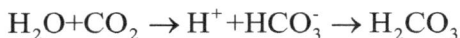

$$H_2O + CO_2 \rightarrow H^+ + HCO_3^- \rightarrow H_2CO_3$$

This is the dominant form of chemical weathering and aides in the breakdown of carbonate minerals like calcite and dolomite and silicate minerals like feldspar. The breakdown of the Na-feldspar, albite, by carbonic acid to form kaolinite clay is as follows:

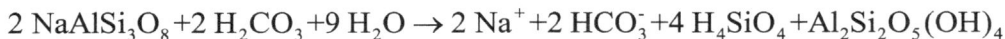

$$2\,NaAlSi_3O_8 + 2\,H_2CO_3 + 9\,H_2O \rightarrow 2\,Na^+ + 2\,HCO_3^- + 4\,H_4SiO_4 + Al_2Si_2O_5(OH)_4$$

Evidence of this reaction in the field would be elevated levels of bicarbonate ($HCO3-$), sodium and silica ions in the water runoff. The breakdown of carbonate minerals:

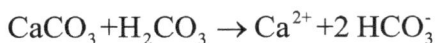

$$CaCO_3 + H_2CO_3 \rightarrow Ca^{2+} + 2\,HCO_3^-$$

The further dissolution of carbonic acid ($H2CO3$) and bicarbonate ($HCO3$) produces $CO2$ gas. Oxidization is also a major contributor to the breakdown of many silicate minerals and formation of secondary minerals (diagenesis) in the early soil profile. Oxidation of Olivine ($FeMgSiO2$) releases Fe, Mg and Si ions. The Mg is soluble in water and is carried in the runoff but the Fe often reacts with oxygen to precipitate $Fe2O3$ (hematite), the oxidized state of iron oxide. Sulfur, a byproduct of decaying organic material will also react to Fe to form pyrite ($FeS2$) but often in reducing environments. Pyrite dissolution leads to high pH levels due to elevated $H+$ ions and further precipitation of $Fe2O3$ ultimately changing the redox conditions of the environment.

Biosphere

Inputs from the biosphere may begin with lichen and other microorganisms that secrete oxalic acid. These microorganisms, associated with the lichen community or independently inhabiting rocks, include a number of blue-green algae, green algae, various fungi, and numerous bacteria. Lichen has long been viewed as the pioneers of soil development as the following statement suggests:

"The initial conversion of rock into soil is carried on by the pioneer lichens and their successors, the mosses, in which the hair-like rhizoids assume the role of roots in breaking down the surface into fine dust"

However, lichens are not necessarily the only pioneering organisms nor the earliest form of soil formation as it has been documented that seed-bearing plants may occupy an area and colonize quicker than lichen. Also, eolian sedimentation can produce high rates of sediment accumulation. Nonetheless, lichen can certainly withstand harsher conditions than most vascular plants and although they have slower colonization rates, do form the dominant group in alpine regions.

Acids released from plant roots include acetic and citric acids. During the decay of organic matter Phenolic acids are released from plant matter and humic and fulvic acids are released by soil microbes. These organic acids speed up chemical weathering by combining with some of the weathering products in a process known as chelation. In the soil profile, the organic acids are often concentrated at the top while carbonic acid plays a larger role towards the bottom or below in the aquifer.

As the soil column develops further into thicker accumulations, larger animals come to inhabit the soil and continue to alter the chemical evolution of their respective niche. Earthworms aerate the soil and convert large amounts of organic matter into rich humus, improving soil fertility. Small burrowing mammals store food, grow young and may hibernate in the pedosphere altering the course of soil evolution. Large mammalian herbivores above ground transport nutrients in form of nitrogen-rich waste and phosphorus-rich antlers while predators leave phosphorus-rich piles of bones on the soil surface, leading the localized enrichment of the soil below.

Redox Conditions in Wetland Soils

Nutrient cycling in lakes and freshwater wetlands depends heavily on redox conditions. Under a few millimeters of water heterotrophic bacteria metabolize and consume oxygen. They therefore deplete the soil of oxygen and create the need for anaerobic respiration. Some anaerobic microbial processes include denitrification, sulfate reduction and methanogenesis and are responsible for the release of N_2 (nitrogen), H_2S (hydrogen sulfide) and CH_4 (methane). Other anaerobic microbial processes are linked to changes in the oxidation state of iron and manganese. As a result of anaerobic decomposition, the soil stores large amounts of organic carbon because decomposition is incomplete.

The redox potential describes which way chemical reactions will proceed in oxygen deficient soils and controls the nutrient cycling in flooded systems. Redox potential, or reduction potential, is used to express the likelihood of an environment to receive electrons and therefore become reduced. For example, if a system already has plenty of electrons (anoxic, organic-rich shale) it is reduced and will likely donate electrons to a part of the system that has a low concentration of electrons, or an oxidized environment, to equilibrate to the chemical gradient. The oxidized environment has high redox potential, whereas the reduced environment has a low redox potential.

The redox potential is controlled by the oxidation state of the chemical species, pH and the amount of oxygen (O_2) there is in the system. The oxidizing environment accepts electrons because of the presence of O_2, which acts as electron acceptors:

$$O_2 + 4\,e^- + 4\,H^+ \rightarrow H_2O$$

This equation will tend to move to the right in acidic conditions which causes higher redox potentials to be found at lower pH levels. Bacteria, heterotrophic organisms, consume oxygen while decomposing organic material which depletes the soils of oxygen, thus increasing the redox potential. In low redox conditions the deposition of ferrous iron ($Fe2+$) will increase with decreasing decomposition rates, thus preserving organic remains and depositing humus. At high redox potential, the oxidized form of iron, ferric iron ($Fe3+$), will be deposited commonly as hematite. By using analytical geochemical tools such as x-ray fluorescence (XRF) or inductively coupled mass spectrometry (ICP-MS) the two forms of Fe ($Fe2+$ and $Fe3+$) can be measured in ancient rocks therefore determining the redox potential for ancient soils.

Such a study was done on Permian through Triassic rocks (300-200 million years old) in Japan and British Columbia. The geologists found hematite throughout the early and middle Permian but began to find the reduced form of iron in pyrite within the ancient soils near the end of the Permian and into the Triassic. This suggests that conditions became less oxygen rich, even anoxic, during the late Permian, which eventually led to the greatest extinction in earth's history, the P-T extinction.

Decomposition in anoxic or reduced soils is also carried out by sulfur-reducing bacteria which, instead of $O2$ use $SO42-$ as an electron acceptor and produce hydrogen sulfide ($H2S$) and carbon dioxide in the process:

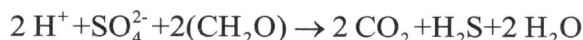

$$2\,H^+ + SO_4^{2-} + 2(CH_2O) \rightarrow 2\,CO_2 + H_2S + 2\,H_2O$$

The H2S gas percolates upwards and reacts with $Fe2+$ and precipitates pyrite, acting as a trap for the toxic H2S gas. However, H2S is still a large fraction of emissions from wetland soils. In most freshwater wetlands there is little sulfate ($SO42-$) so methanogenesis becomes the dominant form of decomposition by methanogenic bacteria only when sulfate is depleted. Acetate, a compound that is a byproduct of fermenting cellulose is split by methanogenic bacteria to produce methane ($CH4$) and carbon dioxide ($CO2$), which are released to the atmosphere. Methane is also released during the reduction of $CO2$ by the same bacteria.

Atmosphere

In the pedosphere it is safe to assume that gases are in equilibrium with the atmosphere. Because plant roots and soil microbes release $CO2$ to the soil, the concentration of bicarbonate($HCO3$) in soil waters is much greater than that in equilibrium with the atmosphere, the high concentration of $CO2$ and the occurrence of metals in soil solutions results in lower pH levels in the soil. Gases that escape from the pedosphere to the atmosphere include the gaseous byproducts of carbonate dissolution, decomposition, redox reactions and microbial photosynthesis. The main inputs from the atmosphere are aeolian sedimentation, rainfall and gas diffusion. Eolian sedimentation includes anything that can be entrained by wind or that stays suspended, seemingly indefinitely, in air and includes a wide variety of aerosol particles, biological particles like pollen and dust to pure quartz sand. Nitrogen is the most abundant constituent in rain (after water), as water vapor utilizes aerosol particles to nucleate rain droplets.

Soil in Forests

Soil is well developed in the forest as suggested by the thick humus layers, rich diversity of large trees and animals that live there. In forests, precipitation exceeds evapotranspiration which results in an excess of water that percolates downward through the soil layers. Slow rates of decomposition leads to large amounts of fulvic acid, greatly enhancing chemical weathering. The downward percolation, in conjunction with chemical weathering leaches magnesium (Mg), iron (Fe), and aluminum (Al) from the soil and transports them downward, a process known as podzolization. This process leads to marked contrasts in the appearance and chemistry of the soil layers.

Soil in the Tropics

Tropical forests (rainforests) receive more insolation and rainfall over longer growing seasons than any other environment on earth. With these elevated temperatures, insolation and rainfall, biomass is extremely productive leading to the production of as much as 800 grams of carbon per square meter per year. Higher temperatures and larger amounts of water contribute to higher rates of chemical weathering. Increased rates of decomposition cause smaller amounts of fulvic acid to percolate and leach metals from the zone of active weathering. Thus, in stark contrast to soil in forests, tropical forests have little to no podzolization and therefore do not have marked visual and chemical contrasts with the soil layers. Instead, the mobile metals Mg, Fe and Al are precipitated as oxide minerals giving the soil a rusty red color.

Soil in Grasslands and Deserts

Precipitation in grasslands is equal to or less than evapotranspiration and causes soil development to operate in relative drought. Leaching and migration of weathering products is therefore decreased. Large amounts of evaporation causes buildup of calcium (Ca) and other large cations flocculate clay minerals and fulvic acids in the upper soil profile. Impermeable clay limits downward percolation of water and fulvic acids, reducing chemical weathering and podzolization. The depth to the maximum concentration of clay increases in areas of increased precipitation and leaching. When leaching is decreased, the Ca precipitates as calcite ($CaCO_3$) in the lower soil levels, a layer known as caliche.

Deserts behave similarly to grasslands but operate in constant drought as precipitation is less than evapotranspiration. Chemical weathering proceeds more slowly than in grasslands and beneath the caliche layer may be a layer of gypsum and halite. To study soils in deserts, pedologists have used the concept of chronosequences to relate timing and development of the soil layers. It has been shown that P is leached very quickly from the system and therefore decreases with increasing age. Furthermore, carbon buildup in the soils is decreased due to slower decomposition rates. As a result, the rates of carbon circulation in the biogeochemical cycle is decreased.

Biosphere

The biosphere is the global sum of all ecosystems. It can also be termed as the zone of life on Earth, a closed system (apart from solar and cosmic radiation and heat from the interior of the Earth), and largely self-regulating. By the most general biophysiological definition, the biosphere is the global ecological system integrating all living beings and their relationships, including their interaction with the elements of the lithosphere, geosphere, hydrosphere, and atmosphere. The biosphere is postulated to have evolved, beginning with a process of biopoesis (life created naturally from non-living matter, such as simple organic compounds) or biogenesis (life created from living matter), at least some 3.5 billion years ago. The earliest evidence for life on Earth includes biogenic graphite found in 3.7 billion-year-old metasedimentary rocks from Western Greenland and microbial mat fossils found in 3.48 billion-year-old sandstone from Western Australia. More recently, in 2015, "remains of biotic life" were found in 4.1 billion-year-old rocks in Western Australia. According to one of the researchers, "If life arose relatively quickly on Earth ... then it could be common in the universe."

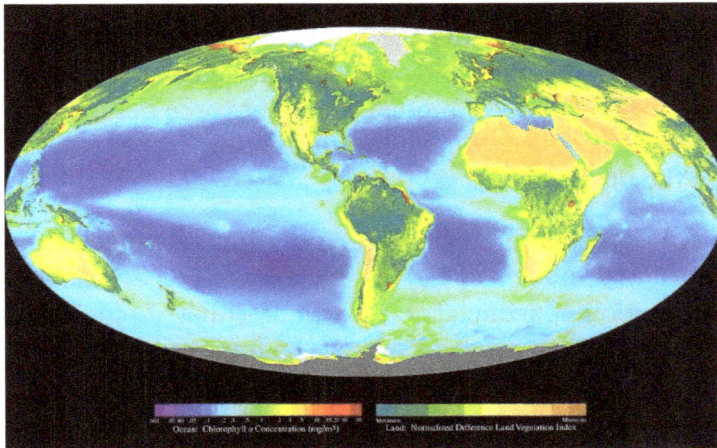

A false-color composite of global oceanic and terrestrial photoautotroph abundance, from September 1997 to August 2000. Provided by the SeaWiFS Project, NASA/Goddard Space Flight Center and ORBIMAGE.

In a general sense, biospheres are any closed, self-regulating systems containing ecosystems. This includes artificial biospheres such as Biosphere 2 and BIOS-3, and potentially ones on other planets or moons.

Origin and use of the Term

A beach scene on Earth, simultaneously showing the lithosphere (ground), hydrosphere (ocean) and atmosphere (air)

The term "biosphere" was coined by geologist Eduard Suess in 1875, which he defined as the place on Earth's surface where life dwells.

While the concept has a geological origin, it is an indication of the effect of both Charles Darwin and Matthew F. Maury on the Earth sciences. The biosphere's ecological context comes from the 1920s preceding the 1935 introduction of the term "ecosystem" by Sir Arthur Tansley. Vernadsky defined ecology as the science of the biosphere. It is an interdisciplinary concept for integrating astronomy, geophysics, meteorology, biogeography, evolution, geology, geochemistry, hydrology and, generally speaking, all life and Earth sciences.

Narrow Definition

Geochemists define the biosphere as being the total sum of living organisms (the "biomass" or "biota" as referred to by biologists and ecologists). In this sense, the biosphere is but one of four separate components of the geochemical model, the other three being *lithosphere, hydrosphere,* and *atmosphere.* The word *ecosphere,* coined during the 1960s, encompasses both biological and physical components of the planet.

The Second International Conference on Closed Life Systems defined *biospherics* as the science and technology of analogs and models of Earth's biosphere; i.e., artificial Earth-like biospheres. Others may include the creation of artificial non-Earth biospheres—for example, human-centered biospheres or a native Martian biosphere—as part of the topic of biospherics.

Extent of Earth's Biosphere

Water covers 71% of the Earth's surface. Image is the Blue Marble photographed from Apollo 17.

Every part of the planet, from the polar ice caps to the equator, features life of some kind. Recent advances in microbiology have demonstrated that microbes live deep beneath the Earth's terrestrial surface, and that the total mass of microbial life in so-called "uninhabitable zones" may, in biomass, exceed all animal and plant life on the surface. The actual thickness of the biosphere on earth is difficult to measure. Birds typically fly at altitudes as high as 1,800 m (5,900 ft; 1.1 mi) and fish live as much as 8,372 m (27,467 ft; 5.202 mi) underwater in the Puerto Rico Trench.

There are more extreme examples for life on the planet: Rüppell's vulture has been found at altitudes of 11,300 m (37,100 ft; 7.0 mi); bar-headed geese migrate at altitudes of at least 8,300 m (27,200 ft; 5.2 mi); yaks live at elevations as high as 5,400 m (17,700 ft; 3.4 mi) above sea level; mountain goats live up to 3,050 m (10,010 ft; 1.90 mi). Herbivorous animals at these elevations depend on lichens, grasses, and herbs.

Microscopic organisms live in every part of the biosphere, including soil, hot springs, "seven miles deep" in the ocean, "40 miles high" in the atmosphere and inside rocks far down within the Earth's crust. Microorganisms, under certain test conditions, have been observed to thrive in the vacuum of outer space. The total amount of soil and subsurface bacterial carbon is estimated as 5 x 1017 g, or the "weight of the United Kingdom". The mass of prokaryote microorganisms — which includes bacteria and archaea, but not the nucleated eukaryote microorganisms — may be as much as 0.8 trillion tons of carbon (of the total biosphere mass, estimated at between 1 and 4 trillion tons). Barophilic marine microbes have been found at more than a depth of 10,000 m (33,000 ft; 6.2 mi) in the Mariana Trench, the deepest spot in the Earth's oceans. In fact, single-celled life forms have been found in the deepest part of the Mariana Trench, by the Challenger Deep, at depths of 11,034 m (36,201 ft; 6.856 mi). Other researchers reported related studies that microorganisms thrive inside rocks up to 580 m (1,900 ft; 0.36 mi) below the sea floor under 2,590 m (8,500 ft; 1.61 mi) of ocean off the coast of the northwestern United States, as well as 2,400 m (7,900 ft; 1.5 mi) beneath the seabed off Japan. Culturable thermophilic microbes have been extracted from cores drilled more than 5,000 m (16,000 ft; 3.1 mi) into the Earth's crust in Sweden, from rocks between 65–75 °C (149–167 °F). Temperature increases with increasing depth into the Earth's crust. The rate at which the temperature increases depends on many factors, including type of crust (continental vs. oceanic), rock type, geographic location, etc. The greatest known temperature at which microbial life can exist is 122 °C (252 °F) (*Methanopyrus kandleri Strain 116), and it is likely that the limit of life in the "deep biosphere" is defined by temperature rather than absolute depth.* On 20 August 2014, scientists confirmed the exis-tence of microorganisms living 800 m (2,600 ft; 0.50 mi) below the ice of Antarctica. According to one researcher, "You can find microbes everywhere — they're extremely adaptable to conditions, and survive wherever they are."

Our biosphere is divided into a number of biomes, inhabited by fairly similar flora and fauna. On land, biomes are separated primarily by latitude. Terrestrial biomes lying within the Arctic and Antarctic Circles are relatively barren of plant and animal life, while most of the more populous biomes lie near the equator.

Specific Biospheres

For this list, if a word is followed by a number, it is usually referring to a specific system or number. Thus:

- Biosphere 1, the planet Earth.
- Biosphere 2, laboratory in Arizona, United States, which contains 3.15 acres (13,000 m2) of closed ecosystem.

- BIOS-3, a closed ecosystem at the Institute of Biophysics in Krasnoyarsk, Siberia, in what was then the Soviet Union.

- Biosphere J (CEEF, Closed Ecology Experiment Facilities), an experiment in Japan.

Extraterrestrial Biospheres

No biospheres have been detected beyond the Earth; therefore, the existence of extraterrestrial biospheres remains hypothetical. The rare Earth hypothesis suggests they should be very rare, save ones composed of microbial life only. On the other hand, Earth analogs may be quite numerous, at least in the Milky Way galaxy. Given limited understanding of abiogenesis, it is currently unknown what percentage of these planets actually develop biospheres.

It is also possible that artificial biospheres will be created during the future, for example on Mars. The process of creating an uncontained system that mimics the function of Earth's biosphere is called terraforming.

Cryosphere

This high resolution image, designed for the Fifth Assessment Report of the IPCC, shows the extent of the regions affected by components of the cryosphere around the world. Over land, continuous permafrost is shown in a dark pink while discontinuous permafrost is shown in a lighter shade of pink. Over much of the northern hemisphere's land area, a semi-transparent white veil depicts the regions that are affected by snowfall at least one day during the period 2000-2012. The bright green line along the southern border of this region shows the maximum snow extent while a black line across the North America, Europe and Asia shows the 50% snow extent line. Glaciers are shown as small golden dots in mountainous areas and in the far northern and southern latitudes. Over the water, ice shelves are shown around Antarctica along with sea ice surrounding the ice shelves. Sea ice is also shown at the North Pole. For both poles. the 30 year average sea ice extent is shown by a yellow outline. In addition, the ice sheets of Greenland and Antarctica are clearly visible.

Overview of the Cryosphere and its larger components, from the UN Environment Programme Global Outlook for Ice and Snow.

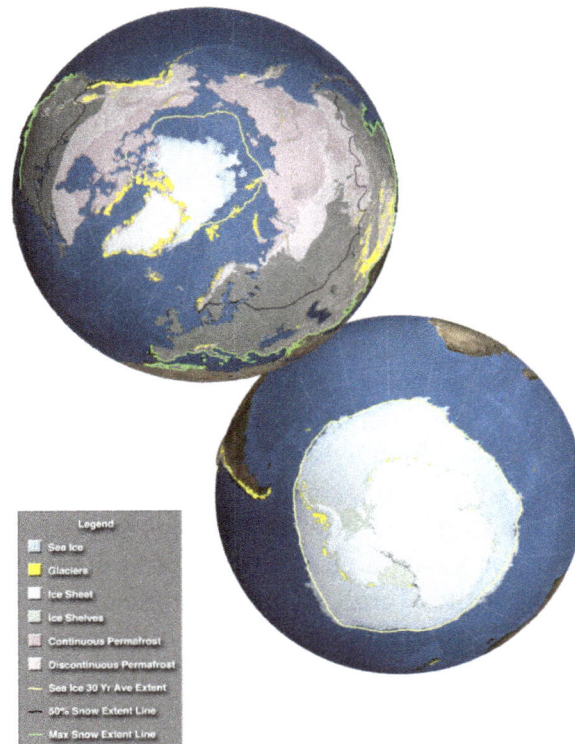

Legend
- Sea Ice
- Glaciers
- Ice Sheet
- Ice Shelves
- Continuous Permafrost
- Discontinuous Permafrost
- Sea Ice 30 Yr Ave Extent
- 50% Snow Extent Line
- Max Snow Extent Line

The cryosphere is those portions of Earth's surface where water is in solid form, including sea ice, lake ice, river ice, snow cover, glaciers, ice caps, ice sheets, and frozen ground (which includes permafrost). Thus, there is a wide overlap with the hydrosphere. The cryosphere is an integral part of the global climate system with important linkages and feedbacks generated through its influence on surface energy and moisture fluxes, clouds, precipitation, hydrology, atmospheric and oceanic circulation. Through these feedback processes, the cryosphere plays a significant role in the global climate and in climate model response to global changes. The term deglaciation describes the retreat of cryo-spheric features. Cryology is the study of cryospheres.

Structure

Frozen water is found on the Earth's surface primarily as snow cover, freshwater ice in lakes and rivers, sea ice, glaciers, ice sheets, and frozen ground and permafrost (permanently frozen ground). The residence time of water in each of these cryospheric sub-systems varies widely. Snow cover and freshwater ice are essentially seasonal, and most sea ice, except for ice in the central Arctic, lasts only a few years if it is not seasonal. A given water particle in glaciers, ice sheets, or ground ice, however, may remain frozen for 10-100,000 years or longer, and deep ice in parts of East Antarctica may have an age approaching 1 million years.

Most of the world's ice volume is in Antarctica, principally in the East Antarctic Ice Sheet. In terms of areal extent, however, Northern Hemisphere winter snow and ice extent comprise the largest area, amounting to an average 23% of hemispheric surface area in January. The large areal extent and the important climatic roles of snow and ice, related to their unique physical properties, indicate that the ability to observe and model snow and ice-cover extent, thickness, and physical properties (radiative and thermal properties) is of particular significance for climate research.

There are several fundamental physical properties of snow and ice that modulate energy exchanges between the surface and the atmosphere. The most important properties are the surface reflectance (albedo), the ability to transfer heat (thermal diffusivity), and the ability to change state (latent heat). These physical properties, together with surface roughness, emissivity, and dielectric characteristics, have important implications for observing snow and ice from space. For example, surface roughness is often the dominant factor determining the strength of radar backscatter . Physical properties such as crystal structure, density, length, and liquid water content are important factors affecting the transfers of heat and water and the scattering of microwave energy.

The surface reflectance of incoming solar radiation is important for the surface energy balance (SEB). It is the ratio of reflected to incident solar radiation, commonly referred to as albedo. Climatologists are primarily interested in albedo integrated over the shortwave portion of the electromagnetic spectrum (~300 to 3500 nm), which coincides with the main solar energy input. Typically, albedo values for non-melting snow-covered surfaces are high (~80-90%) except in the case of forests. The higher albedos for snow and ice cause rapid shifts in surface reflectivity in autumn and spring in high latitudes, but the overall climatic significance of this increase is spatially and temporally modulated by cloud cover. (Planetary albedo is determined principally by cloud cover, and by the small amount of total solar radiation received in high latitudes during winter months.) Summer and autumn are times of high-average cloudiness over the Arctic Ocean so the albedo feedback associated with the large seasonal changes in sea-ice extent is greatly reduced. Groisman *et al. (1994a) observed that snow cover exhibited the greatest influence on the* Earth radiative balance in the spring (April to May) period when incoming solar radiation was greatest over snow-covered areas.

The thermal properties of cryospheric elements also have important climatic consequences. Snow and ice have much lower thermal diffusivities than air. Thermal diffusivity is a measure of the speed at which temperature waves can penetrate a substance. Snow and ice are many orders of magnitude less efficient at diffusing heat than air. Snow cover insulates the ground surface, and sea ice insulates the underlying ocean, decoupling the surface-atmosphere interface with respect to both heat and moisture fluxes. The flux of moisture from a water surface is eliminated by even a thin skin of ice, whereas the flux of heat through thin ice continues to be substantial until it attains a thickness in excess of 30 to 40 cm. However, even a small amount of snow on top of the ice will dramatically reduce the heat flux and slow down the rate of ice growth. The insulating effect of snow also has major implications for the hydrological cycle. In non-permafrost regions, the insulating effect of snow is such that only near-surface ground freezes and deep-water drainage is uninterrupted.

While snow and ice act to insulate the surface from large energy losses in winter, they also act to retard warming in the spring and summer because of the large amount of energy required to melt ice (the latent heat of fusion, 3.34×10^5 J/kg at 0 °C). However, the strong static stability of the atmosphere over areas of extensive snow or ice tends to confine the immediate cooling effect to a relatively shallow layer, so that associated atmospheric anomalies are usually short-lived and local to regional in scale. In some areas of the world such as Eurasia, however, the cooling associated with a heavy snowpack and moist spring soils is known to play a role in modulating the summer monsoon circulation. Gutzler and Preston (1997) recently presented evidence for a similar snow-summer circulation feedback over the southwestern United States.

The role of snow cover in modulating the monsoon is just one example of a short-term cryosphere-climate feedback involving the land surface and the atmosphere. From Figure 1 it can be seen that there are numerous cryosphere-climate feedbacks in the global climate system. These operate over a wide range of spatial and temporal scales from local seasonal cooling of air temperatures to hemispheric-scale variations in ice sheets over time-scales of thousands of years. The feedback mechanisms involved are often complex and incompletely understood. For example, Curry et al. (1995) showed that the so-called "simple" sea ice-albedo feedback involved complex interactions with lead fraction, melt ponds, ice thickness, snow cover, and sea-ice extent.

Snow

Snow cover has the second-largest areal extent of any component of the cryosphere, with a mean maximum areal extent of approximately 47 million km2. Most of the Earth's snow-covered area (SCA) is located in the Northern Hemisphere, and temporal variability is dominated by the seasonal cycle; Northern Hemisphere snow-cover extent ranges from 46.5 million km2 in January to 3.8 million km2 in August. North American winter SCA has exhibited an increasing trend over much of this century (Brown and Goodison 1996; Hughes et al. 1996) largely in response to an increase in precipitation. However, the available satellite data show that the hemispheric winter snow cover has exhibited little interannual variability over the 1972-1996 period, with a coefficient of variation (COV=s.d./mean) for January Northern Hemisphere snow cover of < 0.04. According to Groisman et al. (1994a) Northern Hemisphere spring snow cover should exhibit a decreasing trend to explain an observed increase in Northern Hemisphere spring air temperatures this century. Preliminary estimates of SCA from historical and reconstructed in situ snow-cover data suggest this is the case for Eurasia, but not for North America, where spring snow cover has remained close to current levels over most of this century. Because of the close relationship observed between hemispheric air temperature and snow-cover extent over the period of satellite data (IPCC 1996), there is considerable interest in monitoring Northern Hemisphere snow-cover extent for detecting and monitoring climate change.

Snow cover is an extremely important storage component in the water balance, especially seasonal snowpacks in mountainous areas of the world. Though limited in extent, seasonal snowpacks in the Earth's mountain ranges account for the major source of the runoff for stream flow and groundwater recharge over wide areas of the midlatitudes. For example, over 85% of the annual runoff from the Colorado River basin originates as snowmelt. Snowmelt runoff from the Earth's mountains fills the rivers and recharges the aquifers that over a billion people depend on for their water resources. Further, over 40% of the world's protected areas are in mountains, attesting to their value both as unique ecosystems needing protection and as recreation areas for humans. Climate warming is expected to result in major changes to the partitioning of snow and rainfall, and to the timing of snowmelt, which will have important implications for water use and management. These changes also involve potentially important decadal and longer time-scale feedbacks to the climate system through temporal and spatial changes in soil moisture and runoff to the oceans. (Walsh 1995). Freshwater fluxes from the snow cover into the marine environment may be important, as the total flux is probably of the same magnitude as desalinated ridging and rubble areas of sea ice. In addition, there is an associated pulse of precipitated pollutants which accumulate over the Arctic winter in snowfall and are released into the ocean upon ablation of the sea-ice .

Sea Ice

Sea ice covers much of the polar oceans and forms by freezing of sea water. Satellite data since the early 1970s reveal considerable seasonal, regional, and interannual variability in the sea-ice covers of both hemispheres. Seasonally, sea-ice extent in the Southern Hemisphere varies by a factor of 5, from a minimum of 3-4 million km2 in February to a maximum of 17-20 million km2 in September. The seasonal variation is much less in the Northern Hemisphere where the confined nature and high latitudes of the Arctic Ocean result in a much larger perennial ice cover, and the surrounding land limits the equatorward extent of wintertime ice. Thus, the seasonal variability in Northern Hemisphere ice extent varies by only a factor of 2, from a minimum of 7-9 million km2 in September to a maximum of 14-16 million km2 in March.

The ice cover exhibits much greater regional-scale interannual variability than it does hemispherical. For instance, in the region of the Sea of Okhotsk and Japan, maximum ice extent decreased from 1.3 million km2 in 1983 to 0.85 million km2 in 1984, a decrease of 35%, before rebounding the following year to 1.2 million km2. The regional fluctuations in both hemispheres are such that for any several-year period of the satellite record some regions exhibit decreasing ice coverage while others exhibit increasing ice cover. The overall trend indicated in the passive microwave record from 1978 through mid-1995 shows that the extent of Arctic sea ice is decreasing 2.7% per decade. Subsequent work with the satellite passive-microwave data indicates that from late October 1978 through the end of 1996 the extent of Arctic sea ice decreased by 2.9% per decade while the extent of Antarctic sea ice increased by 1.3% per decade. The Intergovernmental Panel on Climate Change publication *Climate change 2013: The Physical Science Basis stated that sea ice extent for the* Northern Hemisphere showed a decrease of 3.8% ± 0.3% per decade from November 1978 to December 2012.

Lake Ice and River Ice

Ice forms on rivers and lakes in response to seasonal cooling. The sizes of the ice bodies involved are too small to exert other than localized climatic effects. However, the freeze-up/break-up processes respond to large-scale and local weather factors, such that considerable interannual variability exists in the dates of appearance and disappearance of the ice. Long series of lake-ice observations can serve as a proxy climate record, and the monitoring of freeze-up and break-up trends may provide a convenient integrated and seasonally specific index of climatic perturbations. Information on river-ice conditions is less useful as a climatic proxy because ice formation is strongly dependent on river-flow regime, which is affected by precipitation, snow melt, and watershed runoff as well as being subject to human interference that directly modifies channel flow, or that indirectly affects the runoff via land-use practices.

Lake freeze-up depends on the heat storage in the lake and therefore on its depth, the rate and temperature of any inflow, and water-air energy fluxes. Information on lake depth is often unavailable, although some indication of the depth of shallow lakes in the Arctic can be obtained from airborne radar imagery during late winter (Sellman *et al.* 1975) and spaceborne optical imagery during summer (Duguay and Lafleur 1997). The timing of breakup is modified by snow depth on the ice as well as by ice thickness and freshwater inflow.

Frozen Ground and Permafrost

Frozen ground (permafrost and seasonally frozen ground) occupies approximately 54 million km2 of the exposed land areas of the Northern Hemisphere (Zhang et al., 2003) and therefore has the largest areal extent of any component of the cryosphere. Permafrost (perennially frozen ground) may occur where mean annual air temperatures (MAAT) are less than -1 or -2 °C and is generally continuous where MAAT are less than -7 °C. In addition, its extent and thickness are affected by ground moisture content, vegetation cover, winter snow depth, and aspect. The global extent of permafrost is still not completely known, but it underlies approximately 20% of Northern Hemisphere land areas. Thicknesses exceed 600 m along the Arctic coast of northeastern Siberia and Alaska, but, toward the margins, permafrost becomes thinner and horizontally discontinuous. The marginal zones will be more immediately subject to any melting caused by a warming trend. Most of the presently existing permafrost formed during previous colder conditions and is therefore relic. However, permafrost may form under present-day polar climates where glaciers retreat or land emergence exposes unfrozen ground. Washburn (1973) concluded that most continuous permafrost is in balance with the present climate at its upper surface, but changes at the base depend on the present climate and geothermal heat flow; in contrast, most discontinuous permafrost is probably unstable or "in such delicate equilibrium that the slightest climatic or surface change will have drastic disequilibrium effects".

Under warming conditions, the increasing depth of the summer active layer has significant impacts on the hydrologic and geomorphic regimes. Thawing and retreat of permafrost have been reported in the upper Mackenzie Valley and along the southern margin of its occurrence in Manitoba, but such observations are not readily quantified and generalized. Based on average latitudinal gradients of air temperature, an average northward displacement of the southern permafrost boundary by 50-to-150 km could be expected, under equilibrium conditions, for a 1 °C warming.

Only a fraction of the permafrost zone consists of actual ground ice. The remainder (dry permafrost) is simply soil or rock at subfreezing temperatures. The ice volume is generally greatest in the uppermost permafrost layers and mainly comprises pore and segregated ice in Earth material. Measurements of bore-hole temperatures in permafrost can be used as indicators of net changes in temperature regime. Gold and Lachenbruch (1973) infer a 2-4 °C warming over 75 to 100 years at Cape Thompson, Alaska, where the upper 25% of the 400-m thick permafrost is unstable with respect to an equilibrium profile of temperature with depth (for the present mean annual surface temperature of -5 °C). Maritime influences may have biased this estimate, however. At Prudhoe Bay similar data imply a 1.8 °C warming over the last 100 years (Lachenbruch *et al.* 1982). Further complications may be introduced by changes in snow-cover depths and the natural or artificial disturbance of the surface vegetation.

The potential rates of permafrost thawing have been established by Osterkamp (1984) to be two centuries or less for 25-meter-thick permafrost in the discontinuous zone of interior Alaska, assuming warming from -0.4 to 0 °C in 3–4 years, followed by a further 2.6 °C rise. Although the response of permafrost (depth) to temperature change is typically a very slow process (Osterkamp 1984; Koster 1993), there is ample evidence for the fact that the active layer thickness quickly responds to a temperature change (Kane *et al.* 1991). Whether, under a warming or cooling scenario, global climate change will have a significant effect on the duration of frost-free periods in both regions with seasonally and perennially frozen ground.

Glaciers and Ice Sheets

Ice sheets and glaciers are flowing ice masses that rest on solid land. They are controlled by snow accumulation, surface and basal melt, calving into surrounding oceans or lakes and internal dynamics. The latter results from gravity-driven creep flow ("glacial flow") within the ice body and sliding on the underlying land, which leads to thinning and horizontal spreading. Any imbalance of this dynamic equilibrium between mass gain, loss and transport due to flow results in either growing or shrinking ice bodies.

Ice sheets are the greatest potential source of global freshwater, holding approximately 77% of the global total. This corresponds to 80 m of world sea-level equivalent, with Antarctica accounting for 90% of this. Greenland accounts for most of the remaining 10%, with other ice bodies and glaciers accounting for less than 0.5%. Because of their size in relation to annual rates of snow accumulation and melt, the residence time of water in ice sheets can extend to 100,000 or 1 million years. Consequently, any climatic perturbations produce slow responses, occurring over glacial and interglacial periods. Valley glaciers respond rapidly to climatic fluctuations with typical response times of 10–50 years. However, the response of individual glaciers may be asynchronous to the same climatic forcing because of differences in glacier length, elevation, slope, and speed of motion. Oerlemans (1994) provided evidence of coherent global glacier retreat which could be explained by a linear warming trend of 0.66 °C per 100 years.

While glacier variations are likely to have minimal effects upon global climate, their recession may have contributed one third to one half of the observed 20th Century rise in sea level (Meier 1984; IPCC 1996). Furthermore, it is extremely likely that such extensive glacier recession as is currently observed in the Western Cordillera of North America, where runoff from glacierized basins is used for irrigation and hydropower, involves significant hydrological and ecosystem impacts. Effective water-resource planning and impact mitigation in such areas depends upon developing a sophisticated knowledge of the status of glacier ice and the mechanisms that cause it to change. Furthermore, a clear understanding of the mechanisms at work is crucial to interpreting the global-change signals that are contained in the time series of glacier mass balance records.

Combined glacier mass balance estimates of the large ice sheets carry an uncertainty of about 20%. Studies based on estimated snowfall and mass output tend to indicate that the ice sheets are near balance or taking some water out of the oceans. Marinebased studies suggest sea-level rise from the Antarctic or rapid ice-shelf basal melting. Some authors (Paterson 1993; Alley 1997) have suggested that the difference between the observed rate of sea-level rise (roughly 2 mm/y) and the explained rate of sea-level rise from melting of mountain glaciers, thermal expansion of the ocean, etc. (roughly 1 mm/y or less) is similar to the modeled imbalance in the Antarctic (roughly 1 mm/y of sea-level rise; Huybrechts 1990), suggesting a contribution of sea-level rise from the Antarctic.

Relationships between global climate and changes in ice extent are complex. The mass balance of land-based glaciers and ice sheets is determined by the accumulation of snow, mostly in winter, and warm-season ablation due primarily to net radiation and turbulent heat fluxes to melting ice and snow from warm-air advection,(Munro 1990). However, most of Antarctica never experiences surface melting. Where ice masses terminate in the ocean, iceberg calving is the major contributor to mass loss. In this situation, the ice margin may extend out into deep water as a floating ice shelf, such as that in the Ross Sea. Despite the possibility that global warming could result in losses to the Greenland ice sheet being offset by gains to the Antarctic ice sheet, there is major concern about the possibility of a West Antarctic Ice Sheet collapse. The West Antarctic Ice Sheet is grounded on bedrock below sea level, and its collapse has the potential of raising the world sea level 6–7 m over a few hundred years.

Most of the discharge of the West Antarctic Ice Sheet is via the five major ice streams (faster flowing ice) entering the Ross Ice Shelf, the Rutford Ice Stream entering Ronne-Filchner shelf of the Weddell Sea, and the Thwaites Glacier and Pine Island Glacier entering the Amundsen Ice Shelf. Opinions differ as to the present mass balance of these systems (Bentley 1983, 1985), principally because of the limited data. The West Antarctic Ice Sheet is stable so long as the Ross Ice Shelf is constrained by drag along its lateral boundaries and pinned by local grounding.

Geosphere

There are several conflicting definitions for geosphere.

The geosphere may be taken as the collective name for the lithosphere, the hydrosphere, the cryosphere, and the atmosphere.

In Aristotelian physics, the term was applied to four spherical *natural places, concentrically nested around the center of the Earth, as described in the lectures Physica and Meteorologica.* They were believed to explain the motions of the four *terrestrial elements: Earth, Water, Air* and *Fire.*

In modern texts and in Earth system science, geosphere refers to the solid parts of the Earth; it is used along with atmosphere, hydrosphere, and biosphere to describe the systems of the Earth (the interaction of these systems with the magnetosphere is sometimes listed). In that context, sometimes the term lithosphere is used instead of geosphere or solid Earth. The lithosphere, however, only refers to the uppermost layers of the solid Earth (oceanic and continental crustal rocks and uppermost mantle).

Since space exploration began, it has been observed that the extent of the ionosphere or plasmasphere is highly variable, and often much larger than previously appreciated, at times extending to the boundaries of the Earth's magnetosphere or geomagnetosphere. This highly variable outer boundary of *geogenic matter has been referred to as the "geopause,"* to suggest the relative scarcity of such matter beyond it, where the solar wind dominates.

References

- Wallace, John M. and Peter V. Hobbs. Atmospheric Science; An Introductory Survey.Elsevier. Second Edition, 2006. ISBN 978-0-12-732951-2.

- Kennish, Michael J. (2001). Practical handbook of marine science. Marine science series (3rd ed.). CRC Press. p. 35. ISBN 0-8493-2391-6.

- Marq de Villiers (2003). Water: The Fate of Our Most Precious Resource (2 ed.). Toronto, Ontario: McClelland & Stewart. p. 453. ISBN 978-0-7710-2641-6. OCLC 43365804.

- Skinner, B.J. & Porter, S.C.: Physical Geology, page 17, chapt. The Earth: Inside and Out, 1987, John Wiley & Sons, ISBN 0-471-05668-5.

- Campbell, Neil A.; Brad Williamson; Robin J. Heyden (2006). Biology: Exploring Life. Boston, Massachusetts: Pearson Prentice Hall. ISBN 0-13-250882-6.

- Bebarta, Kailash Chandra (2011). Dictionary of Forestry and Wildlife Science. New Delhi: Concept Publishing Company. p. 45. ISBN 978-81-8069-719-7.

- Greve, R.; Blatter, H. (2009). Dynamics of Ice Sheets and Glaciers. Springer. doi:10.1007/978-3-642-03415-2. ISBN 978-3-642-03414-5.

- Wallace, John M. and Peter V. Hobbs. Atmospheric Science; An Introductory Survey.Elsevier. Second Edition, 2006. ISBN 978-0-12-732951-2.

- Borenstein, Seth (19 October 2015). "Hints of life on what was thought to be desolate early Earth". Excite. Yonkers, NY: Mindspark Interactive Network. Associated Press. Retrieved 2015-10-20.

- "Climate Change 2013: The Physical Science Basis" (PDF). ipcc. Intergovernmental Panel on Climate Change. p. 324. Retrieved 16 June 2015.

- "Atmospheric Temperature Trends, 1979-2005 : Image of the Day". Earthobservatory.nasa.gov. 2000-01-01. Retrieved 2014-06-10.

- Pasyanos M. E. (2008-05-15). "Lithospheric Thickness Modeled from Long Period Surface Wave Dispersion" (PDF). Retrieved 2014-04-25.

- University of Georgia (25 August 1998). "First-Ever Scientific Estimate Of Total Bacteria On Earth Shows Far Greater Numbers Than Ever Known Before". Science Daily. Retrieved 10 November 2014.

- Fox, Douglas (20 August 2014). "Lakes under the ice: Antarctica's secret garden". Nature. 512: 244–246. Bibcode:2014Natur.512..244F. doi:10.1038/512244a. Retrieved 21 August 2014.

- Zimmer, Carl (3 October 2013). "Earth's Oxygen: A Mystery Easy to Take for Granted". New York Times. Retrieved 3 October 2013.

- Noffke, Nora; Christian, Daniel; Wacey, David; Hazen, Robert M. (8 November 2013). "Microbially Induced Sedimentary Structures Recording an Ancient Ecosystem in the ca. 3.48 Billion-Year-Old Dresser Formation, Pilbara, Western Australia". Astrobiology (journal). 13 (12): 1103–24. doi:10.1089/ast.2013.1030. PMC 3870916. PMID 24205812. Retrieved 15 November 2013.

Branches of Earth and Planetary Science

Study of the Earth and its various functions such as the change of seasons, tidal movements and seismic activity give us an idea of the structure and nature of life in other planets as well as our own in different eras. This chapter is a compilation of the various branches of Earth and Planetary Science that form an integral part of the broader subject matter.

Atmospheric Sciences

Atmospheric sciences is an umbrella term for the study of the Earth's atmosphere, its processes, the effects other systems have on the atmosphere, and the effects of the atmosphere on these other systems. Meteorology includes atmospheric chemistry and atmospheric physics with a major focus on weather forecasting. Climatology is the study of atmospheric changes (both long and short-term) that define average climates and their change over time, due to both natural and anthropogenic climate variability. Aeronomy is the study of the upper layers of the atmosphere, where dissociation and ionization are important. Atmospheric science has been extended to the field of planetary science and the study of the atmospheres of the planets of the solar system.

Experimental instruments used in atmospheric sciences include satellites, rocketsondes, radiosondes, weather balloons, and lasers.

The term aerology is sometimes used as an alternative term for the study of Earth's atmosphere. Early pioneers in the field include Léon Teisserenc de Bort and Richard Assmann.

Atmospheric Chemistry

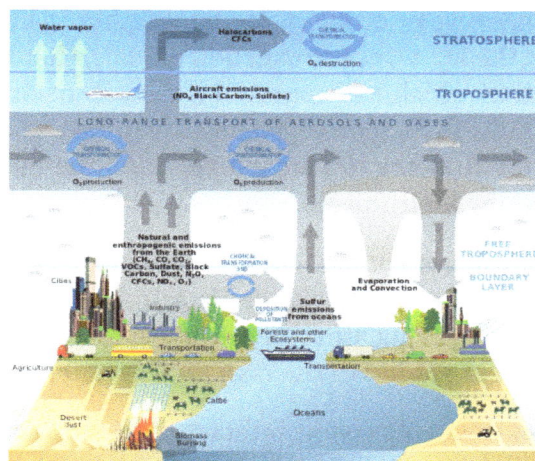

Atmospheric chemistry is a branch of atmospheric science in which the chemistry of the Earth's atmosphere and that of other planets is studied. It is a multidisciplinary field of research and draws on environmental chemistry, physics, meteorology, computer modeling, oceanography, geology and volcanology and other disciplines. Research is increasingly connected with other areas of study such as climatology.

The composition and chemistry of the atmosphere is of importance for several reasons, but primarily because of the interactions between the atmosphere and living organisms. The composition of the Earth's atmosphere has been changed by human activity and some of these changes are harmful to human health, crops and ecosystems. Examples of problems which have been addressed by atmospheric chemistry include acid rain, photochemical smog and global warming. Atmospheric chemistry seeks to understand the causes of these problems, and by obtaining a theoretical understanding of them, allow possible solutions to be tested and the effects of changes in government policy evaluated.

Atmospheric Dynamics

Atmospheric dynamics involves the study of observations and theory dealing with all motion systems of meteorological importance. Common topics studied include diverse phenomena such as thunderstorms, tornadoes, gravity waves, tropical cyclones, extratropical cyclones, jet streams, and global-scale circulations. The goal of dynamical studies is to explain the observed circulations on the basis of fundamental principles from physics. The objectives of such studies incorporate improving weather forecasting, developing methods for predicting seasonal and interannual climate fluctuations, and understanding the implications of human-induced perturbations (e.g., increased carbon dioxide concentrations or depletion of the ozone layer) on the global climate.

Atmospheric Physics

Atmospheric physics is the application of physics to the study of the atmosphere. Atmospheric physicists attempt to model Earth's atmosphere and the atmospheres of the other planets using fluid flow equations, chemical models, radiation balancing, and energy transfer processes in the atmosphere and underlying oceans. In order to model weather systems, atmospheric physicists employ elements of scattering theory, wave propagation models, cloud physics, statistical mechanics and spatial statistics, each of which incorporate high levels of mathematics and physics. Atmospheric physics has close links to meteorology and climatology and also covers the design and construction of instruments for studying the atmosphere and the interpretation of the data they provide, including remote sensing instruments.

In the United Kingdom, atmospheric studies are underpinned by the Meteorological Office. Divisions of the U.S. National Oceanic and Atmospheric Administration (NOAA) oversee research projects and weather modeling involving atmospheric physics. The U.S. National Astronomy and Ionosphere Center also carries out studies of the high atmosphere.

The Earth's magnetic field and the solar wind interact with the atmosphere, creating the ionosphere, Van Allen radiation belts, telluric currents, and radiant energy.

Climatology

WARM EPISODE RELATIONSHIPS   DECEMBER - FEBRUARY

WARM EPISODE RELATIONSHIPS   JUNE - AUGUST

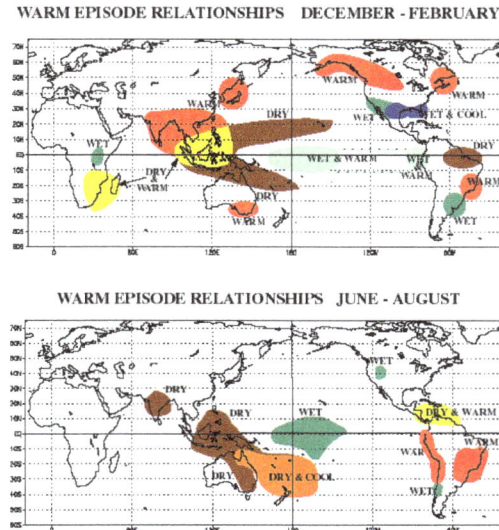

Regional impacts of warm ENSO episodes (El Niño).

In contrast *to* meteorology, which studies short term weather systems lasting up to a few weeks, climatology studies the frequency and trends of those systems. It studies the periodicity of weather events over years to millennia, as well as changes in long-term average weather patterns, in relation to atmospheric conditions. Climatologists, those who practice climatology, study both the nature of climates – local, regional or global – and the natural or human-induced factors that cause climates to change. Climatology considers the past and can help predict future climate change.

Phenomena of climatological interest include the atmospheric boundary layer, circulation patterns, heat transfer (radiative, convective and latent), interactions between the atmosphere and the oceans and land surface (particularly vegetation, land use and topography), and the chemical and physical composition of the atmosphere. Related disciplines include astrophysics, atmospheric physics, chemistry, ecology, physical geography, geology, geophysics, glaciology, hydrology, oceanography, and volcanology.

Atmospheres on Other Celestial Bodies

Earth's Atmosphere

All of the Solar System's planets have atmospheres. This is because their gravity is strong enough to keep gaseous particles close to the surface. Larger gas giants are massive enough to keep large amounts of the light gases hydrogen and helium close by, while the smaller planets lose these gases into space. The composition of the Earth's atmosphere is different from the other planets because the various life processes that have transpired on the planet have introduced free molecular oxygen. Much of Mercury's atmosphere has been blasted away by the solar wind. The only moon that has retained a dense atmosphere is Titan. There is a thin atmosphere on Triton, and a trace of an atmosphere on the Moon.

Planetary atmospheres are affected by the varying degrees of energy received from either the Sun or their interiors, leading to the formation of dynamic weather systems such as hurricanes, (on Earth), planet-wide dust storms (on Mars), an Earth-sized anticyclone on Jupiter (called the Great Red Spot), and holes in the atmosphere (on Neptune). At least one extrasolar planet, HD 189733 b, has been claimed to possess such a weather system, similar to the Great Red Spot but twice as large.

Hot Jupiters have been shown to be losing their atmospheres into space due to stellar radiation, much like the tails of comets. These planets may have vast differences in temperature between their day and night sides which produce supersonic winds, although the day and night sides of HD 189733b appear to have very similar temperatures, indicating that planet's atmosphere effectively redistributes the star's energy around the planet.

Environmental Science

Blue Marble composite images generated by NASA in 2001 (left) and 2002 (right)

Environmental science is an interdisciplinary academic field that integrates physical, biological and information sciences (including ecology, biology, physics, chemistry, zoology, mineralogy, oceanology, limnology, soil science, geology, atmospheric science, and geodesy) to the study of the environment, and the solution of environmental problems. Environmental science emerged from the fields of natural history and medicine during the Enlightenment. Today it provides an integrated, quantitative, and interdisciplinary approach to the study of environmental systems.

Related areas of study include environmental studies and environmental engineering. Environmental studies incorporates more of the social sciences for understanding human relationships, perceptions and policies towards the environment. Environmental engineering focuses on design and technology for improving environmental quality in every aspect.

Environmental scientists work on subjects like the understanding of earth processes, evaluating alternative energy systems, pollution control and mitigation, natural resource management, and the effects of global climate change. Environmental issues almost always include an interaction of physical, chemical, and biological processes. Environmental scientists bring a systems approach to the analysis of environmental problems. Key elements of an effective environmental scientist include the ability to relate space, and time relationships as well as quantitative analysis.

Environmental science came alive as a substantive, active field of scientific investigation in the 1960s and 1970s driven by (a) the need for a multi-disciplinary approach to analyze complex environmental problems, (b) the arrival of substantive environmental laws requiring specific environmental protocols of investigation and (c) the growing public awareness of a need for action in addressing environmental problems. Events that spurred this development included the publication of Rachel Carson's landmark environmental book *Silent Spring* along with major environmental issues becoming very public, such as the 1969 Santa Barbara oil spill, and the Cuyahoga River of Cleveland, Ohio, "catching fire" (also in 1969), and helped increase the visibility of environmental issues and create this new field of study.

Terminology

In common usage, "environmental science" and "ecology" are often used interchangeably, but technically, ecology refers only to the study of organisms and their interactions with each other and their environment. Ecology could be considered a subset of environmental science, which also could involve purely chemical or public health issues (for example) ecologists would be unlikely to study. In practice, there is considerable overlap between the work of ecologists and other environmental scientists.

The National Center for Education Statistics in the United States defines an academic program in environmental science as follows:

A program that focuses on the application of biological, chemical, and physical principles to the study of the physical environment and the solution of environmental problems, including subjects such as abating or controlling environmental pollution and degradation; the interaction between human society and the natural environment; and natural resources management. Includes instruction in biology, chemistry, physics, geosciences, climatology, statistics, and mathematical modeling.

Components

Atmospheric Sciences

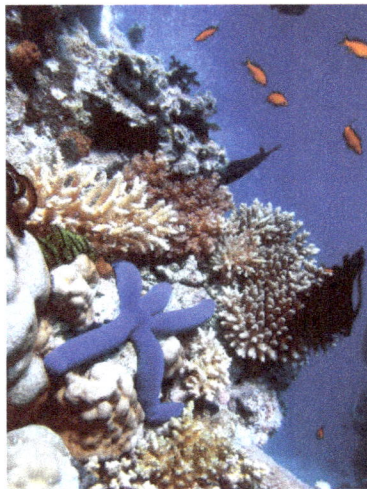

Biodiversity of a coral reef. Corals adapt and modify their environment by forming calcium carbonate skeletons. This provides growing conditions for future generations and forms a habitat for many other species.

Atmospheric sciences focus on the Earth's atmosphere, with an emphasis upon its interrelation to other systems. Atmospheric sciences can include studies of meteorology, greenhouse gas phenomena, atmospheric dispersion modeling of airborne contaminants, sound propagation phenomena related to noise pollution, and even light pollution.

Taking the example of the global warming phenomena, physicists create computer models of atmospheric circulation and infra-red radiation transmission, chemists examine the inventory of atmospheric chemicals and their reactions, biologists analyze the plant and animal contributions to carbon dioxide fluxes, and specialists such as meteorologists and oceanographers add additional breadth in understanding the atmospheric dynamics.

Ecology

Ecology is the study of the interactions between organisms and their environment. Ecologists might investigate the relationship between a population of organisms and some physical characteristic of their environment, such as concentration of a chemical; or they might investigate the interaction between two populations of different organisms through some symbiotic or competitive relationship.

For example, an interdisciplinary analysis of an ecological system which is being impacted by one or more stressors might include several related environmental science fields. In an estuarine setting where a proposed industrial development could impact certain species by water and air pollution, biologists would describe the flora and fauna, chemists would analyze the transport of water pollutants to the marsh, physicists would calculate air pollution emissions and geologists would assist in understanding the marsh soils and bay muds.

Environmental Chemistry

Environmental chemistry is the study of chemical alterations in the environment. Principal areas of study include soil contamination and water pollution. The topics of analysis include chemical degradation in the environment, multi-phase transport of chemicals (for example, evaporation of a solvent containing lake to yield solvent as an air pollutant), and chemical effects upon biota.

As an example study, consider the case of a leaking solvent tank which has entered the habitat soil of an endangered species of amphibian. As a method to resolve or understand the extent of soil contamination and subsurface transport of solvent, a computer model would be implemented. Chemists would then characterize the molecular bonding of the solvent to the specific soil type, and biologists would study the impacts upon soil arthropods, plants, and ultimately pond-dwelling organisms that are the food of the endangered amphibian.

Geosciences

Geosciences include environmental geology, environmental soil science, volcanic phenomena and evolution of the Earth's crust. In some classification systems this can also include hydrology, including oceanography.

As an example study of soils erosion, calculations would be made of surface runoff by soil scientists. Fluvial geomorphologists would assist in examining sediment transport in overland flow.

Physicists would contribute by assessing the changes in light transmission in the receiving waters. Biologists would analyze subsequent impacts to aquatic flora and fauna from increases in water turbidity.

Open-pit coal mining at Garzweiler, Germany

Regulations Driving the Studies

Environmental science examines the effects of humans on nature (Glen Canyon Dam in the U.S.)

In the U.S. the National Environmental Policy Act (NEPA) of 1969 set forth requirements for analysis of major projects in terms of specific environmental criteria. Numerous state laws have echoed these mandates, applying the principles to local-scale actions. The upshot has been an explosion of documentation and study of environmental consequences before the fact of development actions.

One can examine the specifics of environmental science by reading examples of Environmental Impact Statements prepared under NEPA such as: *Wastewater treatment expansion options discharging into the San Diego/Tijuana Estuary, Expansion of the San Francisco International Airport, Development of the Houston, Metro Transportation system, Expansion of the metropolitan Boston MBTA transit system, and Construction of Interstate 66 through Arlington, Virginia.*

In England and Wales the Environment Agency (EA), formed in 1996, is a public body for protecting and improving the environment and enforces the regulations listed on the communities and local government site. (formerly the office of the deputy prime minister). The agency was set up under the Environment Act 1995 as an independent body and works closely with UK Government to enforce the regulations.

Soil Science

Soil science is the study of soil as a natural resource on the surface of the Earth including soil formation, classification and mapping; physical, chemical, biological, and fertility properties of soils; and these properties in relation to the use and management of soils.

A sylviculturist, at work

Sometimes terms which refer to branches of soil science, such as pedology (formation, chemistry, morphology and classification of soil) and edaphology (influence of soil on organisms, especially plants), are used as if synonymous with soil science. The diversity of names associated with this discipline is related to the various associations concerned. Indeed, engineers, agronomists, chemists, geologists, physical geographers, ecologists, biologists, microbiologists, sylviculturists, sanitarians, archaeologists, and specialists in regional planning, all contribute to further knowledge of soils and the advancement of the soil sciences.

Soil scientists have raised concerns about how to preserve soil and arable land in a world with a growing population, possible future water crisis, increasing per capita food consumption, and land degradation.

Fields of Study

Soil occupies the pedosphere, one of Earth's spheres that the geosciences use to organize the Earth conceptually. This is the conceptual perspective of pedology and edaphology, the two main branches of soil science. Pedology is the study of soil in its natural setting. Edaphology is the study of soil in relation to soil-dependent uses. Both branches apply a combination of soil physics, soil chemistry, and soil biology. Due to the numerous interactions between the biosphere, atmosphere and hydrosphere that are hosted within the pedosphere, more integrated, less soil-centric concepts are also valuable. Many concepts essential to understanding soil come from individuals not identifiable strictly as soil scientists. This highlights the interdisciplinary nature of soil concepts.

Research

Dependence on and curiosity about soil, exploring the diversity and dynamics of this resource continues to yield fresh discoveries and insights. New avenues of soil research are compelled by a need to understand soil in the context of climate change, greenhouse gases, and carbon sequestration. Interest in maintaining the planet's biodiversity and in exploring past cultures has also stimulated renewed interest in achieving a more refined understanding of soil.

Mapping

Most empirical knowledge of soil in nature comes from soil survey efforts. Soil survey, or soil mapping, is the process of determining the soil types or other properties of the soil cover over a landscape, and mapping them for others to understand and use. It relies heavily on distinguishing the individual influences of the five classic soil forming factors. This effort draws upon geomorphology, physical geography, and analysis of vegetation and land-use patterns. Primary data for the soil survey are acquired by field sampling and supported by remote sensing.

Classification

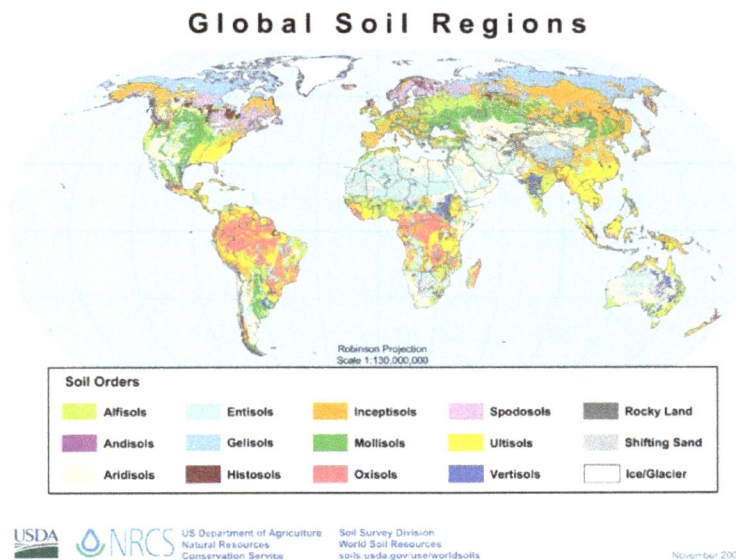

Map of global soil regions from the USDA

As of 2006, the World Reference Base for Soil Resources, via its Land & Water Development division, is the pre-eminent soil classification system. It replaces the previous FAO soil classification.

The WRB borrows from modern soil classification concepts, including USDA soil taxonomy. The classification is based mainly on soil morphology as an expression pedogenesis. A major difference with USDA soil taxonomy is that soil climate is not part of the system, except insofar as climate influences soil profile characteristics.

Many other classification schemes exist, including vernacular systems. The structure in vernacular systems are either nominal, giving unique names to soils or landscapes, or descriptive, naming soils by their characteristics such as red, hot, fat, or sandy. Soils are distinguished by obvious characteristics, such as physical appearance (e.g., color, texture, landscape position), performance (e.g., production capability, flooding), and accompanying vegetation. A vernacular distinction familiar to many is classifying texture as heavy or light. Light soil content and better structure, take less effort to turn and cultivate. Contrary to popular belief, light soils do not weigh less than heavy soils on an air dry basis nor do they have more porosity.

History

Vasily Dokuchaev, a Russian geologist, geographer and early soil scientist, is credited with identifying soil as a resource whose distinctness and complexity deserved to be separated conceptually from geology and crop production and treated as a whole.

Previously, soil had been considered a product of chemical transformations of rocks, a dead substrate from which plants derive nutritious elements. Soil and bedrock were in fact equated. Dokuchaev considers the soil as a natural body having its own genesis and its own history of development, a body with complex and multiform processes taking place within it. The soil is considered as different from bedrock. The latter becomes soil under the influence of a series of soil-formation factors (climate, vegetation, country, relief and age). According to him, soil should be called the "daily" or outward horizons of rocks regardless of the type; they are changed naturally by the common effect of water, air and various kinds of living and dead organisms.

A 1914 encyclopedic definition: "the different forms of earth on the surface of the rocks, formed by the breaking down or weathering of rocks". serves to illustrate the historic view of soil which persisted from the 19th century. Dokuchaev's late 19th century soil concept developed in the 20th century to one of soil as earthy material that has been altered by living processes. A corollary concept is that soil without a living component is simply a part of earth's outer layer.

Further refinement of the soil concept is occurring in view of an appreciation of energy transport and transformation within soil. The term is popularly applied to the material on the surface of the Earth's moon and Mars, a usage acceptable within a portion of the scientific community. Accurate to this modern understanding of soil is Nikiforoff's 1959 definition of soil as the "excited skin of the sub aerial part of the earth's crust".

Areas of Practice

Academically, soil scientists tend to be drawn to one of five areas of specialization: microbiology, pedology, edaphology, physics or chemistry. Yet the work specifics are very much dictated by the challenges facing our civilization's desire to sustain the land that supports it, and the distinctions between the sub-disciplines of soil science often blur in the process. Soil science professionals commonly stay current in soil chemistry, soil physics, soil microbiology, pedology, and applied soil science in related disciplines

One interesting effort drawing in soil scientists in the USA as of 2004[update] is the Soil Quality Initiative. Central to the Soil Quality Initiative is developing indices of soil health and then monitoring them in a way that gives us long term (decade-to-decade) feedback on our performance as stewards of the planet. The effort includes understanding the functions of soil microbiotic crusts and exploring the potential to sequester atmospheric carbon in soil organic matter. The concept of soil quality, however, has not been without its share of controversy and criticism, including critiques by Nobel Laureate Norman Borlaug and World Food Prize Winner Pedro Sanchez.

A more traditional role for soil scientists has been to map soils. Most every area in the United States now has a published soil survey, which includes interpretive tables as to how soil properties support or limit activities and uses. An internationally accepted soil taxonomy allows uniform communication of soil characteristics and functions. National and international soil survey efforts have given the profession unique insights into landscape scale functions. The landscape functions that soil scientists are called upon to address in the field seem to fall roughly into six areas:

- Land-based treatment of wastes
    - Septic system
    - Manure
    - Municipal biosolids
    - Food and fiber processing waste
- Identification and protection of environmentally critical areas
    - Sensitive and unstable soils
    - Wetlands
    - Unique soil situations that support valuable habitat, and ecosystem diversity
- Management for optimum land productivity
    - Silviculture
    - Agronomy
        - Nutrient management
        - Water management
    - Native vegetation
    - Grazing
- Management for optimum water quality
    - Stormwater management
    - Sediment and erosion control
- Remediation and restoration of damaged lands
    - Mine reclamation
    - Flood and storm damage
    - Contamination

- Sustainability of desired uses
  - o Soil conservation

There are also practical applications of soil science that might not be apparent from looking at a published soil survey.

- Radiometric dating: specifically a knowledge of local pedology is used to date prior activity at the site
  - o Stratification (archeology) where soil formation processes and preservative qualities can inform the study of archaeological sites
  - o Geological phenomena
    - Landslides
    - Active faults
- Altering soils to achieve new uses
  - o Vitrification to contain radioactive wastes
  - o Enhancing soil microbial capabilities in degrading contaminants (bioremediation).
  - o Carbon sequestration
  - o Environmental soil science
- Pedology
  - o Soil genesis
  - o Pedometrics
  - o Soil morphology
    - Soil micromorphology
  - o Soil classification
    - USDA soil taxonomy
- Soil biology
  - o Soil microbiology
- Soil chemistry
  - o Soil biochemistry
  - o Soil mineralogy
- Soil physics
  - o Pedotransfer function

- o   Soil mechanics and engineering
- Soil hydrology, hydropedology

Fields of Application in Soil Science

- Climate change
- Ecosystem studies
- Pedotransfer function
- Soil fertility / Nutrient management
- Soil management
- Soil survey
- Standard methods of analysis
- Watershed and wetland studies

Related Disciplines

- Agricultural sciences
    - o   Agricultural soil science
    - o   Agrophysics science
    - o   Irrigation management
- Anthropology
    - o   archaeological stratigraphy
- Environmental science
    - o   Landscape ecology
- Physical geography
    - o   Geomorphology
- Geology
    - o   Biogeochemistry
    - o   Geomicrobiology
- Hydrology
    - o   Hydrogeology
- Waste management
- Wetland science

Oceanography

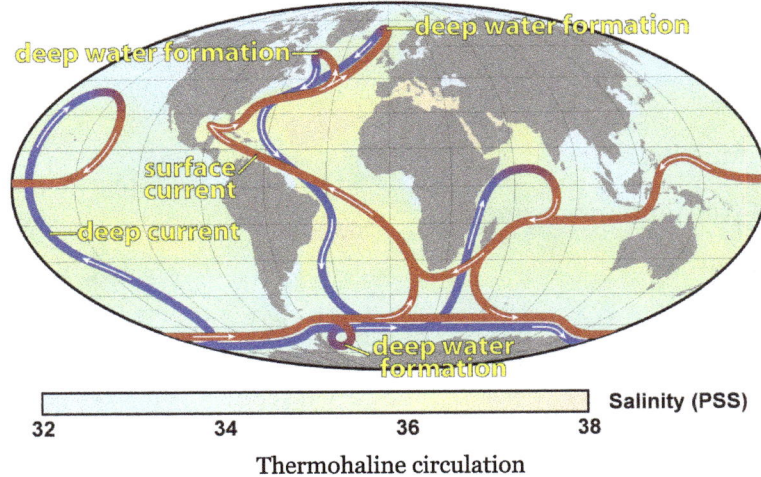

Thermohaline circulation

Oceanography also known as oceanology, is the branch of Earth science that studies the ocean. It covers a wide range of topics, including ecosystem dynamics; ocean currents, waves, and geophysical fluid dynamics; plate tectonics and the geology of the sea floor; and fluxes of various chemical substances and physical properties within the ocean and across its boundaries. These diverse topics reflect multiple disciplines that oceanographers blend to further knowledge of the world ocean and understanding of processes within: astronomy, biology, chemistry, climatology, geography, geology, hydrology, meteorology and physics. Paleoceanography studies the history of the oceans in the geologic past.

History

Map of the Gulf Stream by Benjamin Franklin, 1769-1770. Courtesy of the NOAA Photo Library.

Early History

Humans first acquired knowledge of the waves and currents of the seas and oceans in pre-historic times. Observations on tides were recorded by Aristotle and Strabo. Early exploration of the oceans was primarily for cartography and mainly limited to its surfaces and of the animals that fishermen brought up in nets, though depth soundings by lead line were taken.

Although Juan Ponce de León in 1513 first identified the Gulf Stream, and the current was well-known to mariners, Benjamin Franklin made the first scientific study of it and gave it its name. Franklin measured water temperatures during several Atlantic crossings and correctly explained the Gulf Stream's cause. Franklin and Timothy Folger printed the first map of the Gulf Stream in 1769-1770.

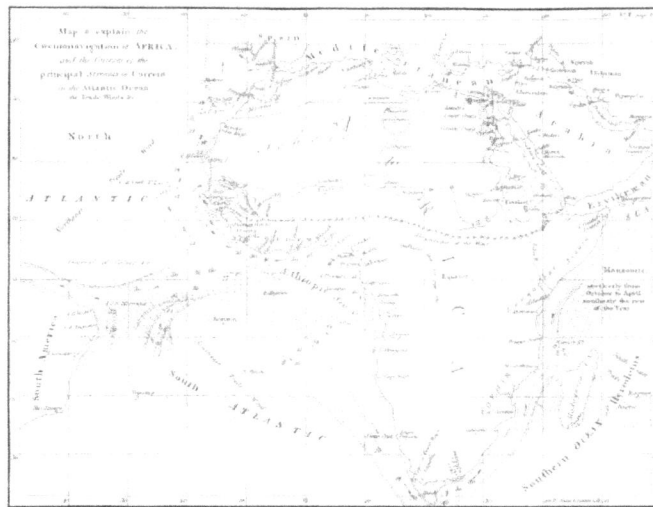

1799 map of the currents in the Atlantic and Indian Oceans, by James Rennell

Information on the currents of the Pacific Ocean was gathered by explorers of the late 18th century, including James Cook and Louis Antoine de Bougainville. James Rennell wrote the first scientific textbooks on oceanography, detailing the current flows of the Atlantic and Indian oceans. During a voyage around the Cape of Good Hope in 1777, he mapped *"the banks and currents at the Lagullas"*. He was also the first to understand the nature of the intermittent current near the Isles of Scilly, (now known as Rennell's Current).

Sir James Clark Ross took the first modern sounding in deep sea in 1840, and Charles Darwin published a paper on reefs and the formation of atolls as a result of the Second voyage of HMS Beagle in 1831-6. Robert FitzRoy published a four-volume report of the Beagle's three voyages. In 1841–1842 Edward Forbes undertook dredging in the Aegean Sea that founded marine ecology.

The first superintendent of the United States Naval Observatory (1842–1861), Matthew Fontaine Maury devoted his time to the study of marine meteorology, navigation, and charting prevailing winds and currents. His 1855 textbook *Physical Geography of the Sea* was one of the first comprehensive oceanography studies. Many nations sent oceanographic observations to Maury at the Naval Observatory, where he and his colleagues evaluated the information and distributed the results worldwide.

Modern Oceanography

Despite all this, human knowledge of the oceans remained confined to the topmost few fathoms of the water and a small amount of the bottom, mainly in shallow areas. Almost nothing was known of the ocean depths. The Royal Navy's efforts to chart all of the world's coastlines in the mid-19th century reinforced the vague idea that most of the ocean was very deep, although little more was known. As exploration ignited both popular and scientific interest in the polar regions and Africa, so too did the mysteries of the unexplored oceans.

H.M.S. CHALLENGER UNDER SAIL, 1874.

HMS *Challenger undertook the first global marine research expedition in 1872.*

The seminal event in the founding of the modern science of oceanography was the 1872-76 Challenger expedition. As the first true oceanographic cruise, this expedition laid the groundwork for an entire academic and research discipline. In response to a recommendation from the Royal Society, The British Government announced in 1871 an expedition to explore world's oceans and conduct appropriate scientific investigation. Charles Wyville Thompson and Sir John Murray launched the Challenger expedition. The Challenger, leased from the Royal Navy, was modified for scientific work and equipped with separate laboratories for natural history and chemistry. Under the scientific supervision of Thomson, Challenger travelled nearly 70,000 nautical miles (130,000 km) surveying and exploring. On her journey circumnavigating the globe, 492 deep sea soundings, 133 bottom dredges, 151 open water trawls and 263 serial water temperature observations were taken. Around 4,700 new species of marine life were discovered. The result was the *Report Of The Scientific Results of the Exploring Voyage of H.M.S. Challenger during the years 1873-76*. Murray, who supervised the publication, described the report as "the greatest advance in the knowledge of our planet since the celebrated discoveries of the fifteenth and sixteenth centuries". He went on to found the academic discipline of oceanography at the University of Edinburgh, which remained the centre for oceanographic research well into the 20th century. Murray was the first to study marine trenches and in particular the Mid-Atlantic Ridge, and map the sedimentary deposits in the oceans. He tried to map out the world's ocean currents based on salinity and temperature observations, and was the first to correctly understand the nature of coral reef development.

In the late 19th century, other Western nations also sent out scientific expeditions (as did private individuals and institutions). The first purpose built oceanographic ship, the Albatros, was built in 1882. In 1893, Fridtjof Nansen allowed his ship, Fram, to be frozen in the Arctic ice. This enabled him to obtain oceanographic, meteorological and astronomical data at a stationary spot over an extended period.

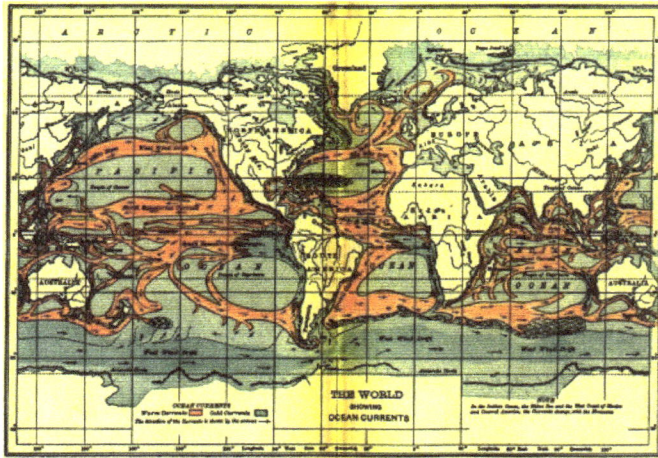

Ocean currents (1911)

Between 1907 and 1911 Otto Krümmel published the *Handbuch der Ozeanographie,* which became influential in awakening public interest in oceanography. The four-month 1910 North Atlantic expedition headed by John Murray and Johan Hjort was the most ambitious research oceanographic and marine zoological project ever mounted until then, and led to the classic 1912 book *The Depths of the Ocean.*

The first acoustic measurement of sea depth was made in 1914. Between 1925 and 1927 the "Meteor" expedition gathered 70,000 ocean depth measurements using an echo sounder, surveying the Mid-Atlantic ridge.

Sverdrup, Johnson and Fleming published *The Oceans* in 1942, which was a major landmark. *The Sea* (in three volumes, covering physical oceanography, seawater and geology) edited by M.N. Hill was published in 1962, while Rhodes Fairbridge's *Encyclopedia of Oceanography was published in 1966.*

The Great Global Rift, running along the Mid Atlantic Ridge, was discovered by Maurice Ewing and Bruce Heezen in 1953; in 1954 a mountain range under the Arctic Ocean was found by the Arctic Institute of the USSR. The theory of seafloor spreading was developed in 1960 by Harry Hammond Hess. The Ocean Drilling Program started in 1966. Deep sea vents were discovered in 1977 by John Corlis and Robert Ballard in the submersible DSV *Alvin.*

In the 1950s, Auguste Piccard invented the bathyscaphe and used the Trieste to investigate the ocean's depths. The United States nuclear submarine Nautilus made the first journey under the ice to the North Pole in 1958. In 1962 the FLIP (Floating Instrument Platform), a 355-foot spar buoy, was first deployed.

From the 1970s, there has been much emphasis on the application of large scale computers to oceanography to allow numerical predictions of ocean conditions and as a part of overall environmental change prediction. An oceanographic buoy array was established in the Pacific to allow prediction of El Niño events.

1990 saw the start of the World Ocean Circulation Experiment (WOCE) which continued until 2002. Geosat seafloor mapping data became available in 1995.

In recent years studies advanced particular knowledge on ocean acidification, ocean heat content, ocean currents, the El Niño phenomenon, mapping of methane hydrate deposits, the carbon cycle, coastal erosion, weathering and climate feedbacks in regards to climate change interactions.

Study of the oceans is linked to understanding global climate changes, potential global warming and related biosphere concerns. The atmosphere and ocean are linked because of evaporation and precipitation as well as thermal flux (and solar insolation). Wind stress is a major driver of ocean currents while the ocean is a sink for atmospheric carbon dioxide. All these factors relate to the ocean's biogeochemical setup.

Branches

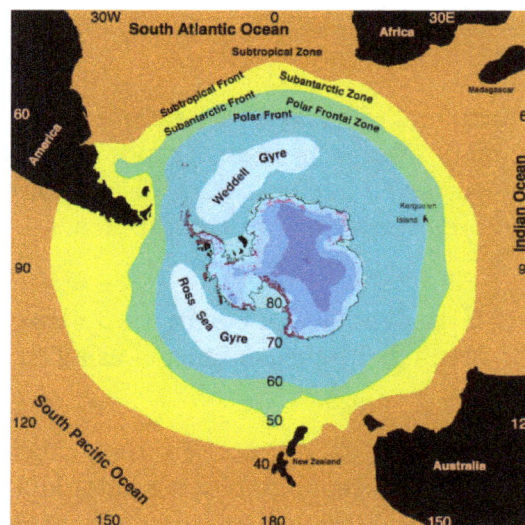

Oceanographic frontal systems on the Southern Hemisphere

The study of oceanography is divided into these four branches:

- Biological oceanography, or marine biology, investigates the ecology of marine organisms in the context of the physical, chemical, and geological characteristics of their ocean environment and the biology of individual marine organisms.

- Chemical oceanography, or marine chemistry, is the study of the chemistry of the ocean and its chemical constraint programming is a programming paradigm wherein relations between variables are stated in the form of constraints. l interaction with the atmosphere.

- Geological oceanography, or marine geology, is the study of the geology of the ocean floor including plate tectonics and paleoceanography.

- Physical oceanography, or marine physics, studies the ocean's physical attributes including temperature-salinity structure, mixing, surface waves, internal waves, surface tides, internal tides, and currents.

Ocean Acidification

Ocean acidification describes the decrease in ocean pH that is caused by anthropogenic carbon dioxide (CO_2) emissions into the atmosphere. Seawater is slightly alkaline and had a preindustrial pH of about 8.2. More recently, anthropogenic activities have steadily increased the carbon dioxide content of the atmosphere; about 30–40% of the added CO_2 is absorbed by the oceans, forming carbonic acid and lowering the pH (now below 8.1) through ocean acidification. The pH is expected to reach 7.7 by the year 2100.

An important element for the skeletons of marine animals is calcium, but calcium carbonate becomes more soluble with pressure, so carbonate shells and skeletons dissolve below the carbonate compensation depth. Calcium carbonate becomes more soluble at lower pH, so ocean acidification is likely to affect marine organisms with calcareous shells, such as oysters, clams, sea urchins and corals, and the carbonate compensation depth will rise closer to the sea surface. Affected planktonic organisms will include pteropods, coccolithophorids and foraminifera, all important in the food chain. In tropical regions, corals are likely to be severely affected as they become less able to build their calcium carbonate skeletons, in turn adversely impacting other reef dwellers.

The current rate of ocean chemistry change seems to be unprecedented in Earth's geological history, making it unclear how well marine ecosystems will adapt to the shifting conditions of the near future. Of particular concern is the manner in which the combination of acidification with the expected additional stressors of higher temperatures and lower oxygen levels will impact the seas.

Ocean Currents

Since the early ocean expeditions in oceanography, a major interest was the study of the ocean currents and temperature measurements. The tides, the Coriolis effect, changes in direction and strength of wind, salinity and temperature are the main factors determining ocean currents. The thermohaline circulation (THC) *thermo- referring to* temperature and *-haline referring to* salt content connects 4 of 5 ocean basins and is primarily dependent on the density of sea water. Ocean currents such as the Gulf Stream are wind-driven surface currents.

Ocean Heat Content

Oceanic heat content (OHC) refers to the heat stored in the ocean. The changes in the ocean heat play an important role in sea level rise, because of thermal expansion. Ocean warming accounts for 90% of the energy accumulation from global warming between 1971 and 2010.

Oceanographic Institutions

The first international organization of oceanography was created in 1902 as the International Council for the Exploration of the Sea. In 1903 the Scripps Institution of Oceanography was founded, followed by Woods Hole Oceanographic Institution in 1930, Virginia Institute of Marine Science in 1938, and later the Lamont-Doherty Earth Observatory at Columbia University, and the

School of Oceanography at University of Washington. In Britain, the National Oceanography Centre (an institute of the Natural Environment Research Council) is the successor to the UK's Institute of Oceanographic Sciences. In Australia, CSIRO Marine and Atmospheric Research (CMAR), is a leading centre. In 1921 the International Hydrographic Bureau (IHB) was formed in Monaco.

Oceanographic Museum

Related Disciplines

- Biogeochemistry
- Biogeography
- Climatology
- Coastal geography
- Environmental science
- Geophysics
- Glaciology
- Hydrography
- Hydrology
- Limnology
- Meteorology
- MetOcean
- Marine science

Geography

Geography is a field of science devoted to the study of the lands, the features, the inhabitants, and the phenomena of Earth.

The first person to use the word was Eratosthenes (276–194 BC). Four historical traditions in geographical research are spatial analysis of the natural and the human phenomena (geography as the study of distribution), area studies (places and regions), study of the human-land relationship, and research in the Earth sciences. Nonetheless, modern geography is an all-encom-passing discipline that foremost seeks to understand the Earth and all of its human and natural complexities—not merely where objects are, but how they have changed and come to be. Geography has been called "the world discipline" and "the bridge between the human and the physical sci-ence". Geography is divided into two main branches: human geography and physical geography.

Physical map of the Earth with political borders as of 2004

Introduction

Traditionally, geographers have been viewed the same way as cartographers and people who study place names and numbers. Although many geographers are trained in toponymy and cartology, this is not their main preoccupation. Geographers study the space and the temporal database distribution of phenomena, processes, and features as well as the interaction of humans and their environment. Because space and place affect a variety of topics, such as economics, health, climate, plants and animals; geography is highly interdisciplinary. The interdisciplinary nature of the geographical approach depends on an attentiveness to the relationship between physical and human phenomena and its spatial patterns.

Names of places...are not geography...know by heart a whole gazetteer full of them would not, in itself, constitute anyone a geographer. Geography has higher aims than this: it seeks to classify phenomena (alike of the natural and of the political world, in so far as it treats of the latter), to compare, to generalize, to ascend from effects to causes, and, in doing so, to trace out the laws of nature and to mark their influences upon man. This is 'a description of the world'—that is Geography. In a word Geography is a Science—a thing not of mere names but of argument and reason, of cause and effect.

— *William Hughes, 1863*

Just as all phenomena exist in time and thus have a history, they also exist in space and have a geography.

— *United States National Research Council, 1997*

Geography as a discipline can be split broadly into two main subsidiary fields: human geography and physical geography. The former largely focuses on the built environment and how humans create, view, manage, and influence space. The latter examines the natural environment, and how organisms, climate, soil, water, and landforms produce and interact. The difference between these approaches led to a third field, environmental geography, which combines the physical and the human geography, and looks at the interactions between the environment and humans.

Branches

Physical Geography

Physical geography (or physiography) focuses on geography as an Earth science. It aims to understand the physical problems and the issues of lithosphere, hydrosphere, atmosphere, pedosphere, and global flora and fauna patterns (biosphere).

- Physical geography can be divided into many broad categories, including:

Biogeography

Climatology & meteorology

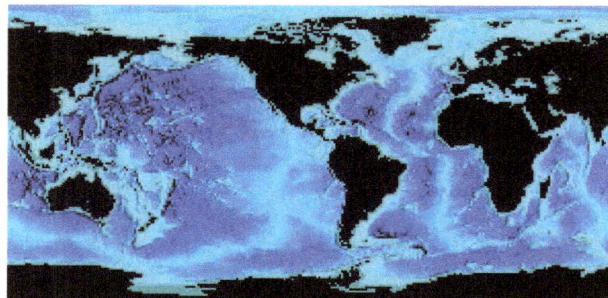

Oceanography

Human Geography

Human geography is a branch of geography that focuses on the study of patterns and processes that shape the human society. It encompasses the human, political, cultural, social, and economic aspects.

- Human geography can be divided into many broad categories, such as:

Cultural geography

Religion geography

Social geography

Urban geography

Various approaches to the study of human geography have also arisen through time and include:

- Behavioral geography

- Feminist geography

- Culture theory

- Geosophy

Integrated Geography

Integrated geography is the branch of geography that describes the spatial aspects of interactions between humans and the natural world. It requires an understanding of the traditional aspects of the physical and the human geography, as well as the ways that human societies conceptualize the environment.

Integrated geography has emerged as a bridge between the human and the physical geography, as a result of the increasing specialisation of the two sub-fields. Furthermore, as human relationship with the environment has changed as a result of globalization and technological change, a new approach was needed to understand the changing and dynamic relationship. Examples of areas of research in the environmental geography include: emergency management, environmental management, sustainability, and political ecology.

Geomatics

Geomatics is a branch of geography that has emerged since the quantitative revolution in geography in the mid-1950s. Geomatics involves the use of traditional spatial techniques used in cartography and topography and their application to computers. Geomatics has become a widespread field with many other disciplines, using techniques such as GIS and remote sensing. Geomatics has also led to a revitalization of some geography departments, especially in Northern America where the subject had a declining status during the 1950s.

Geomatics encompasses a large area of fields involved with spatial analysis, such as Cartography, Geographic information systems (GIS), Remote sensing, and Global positioning systems (GPS).

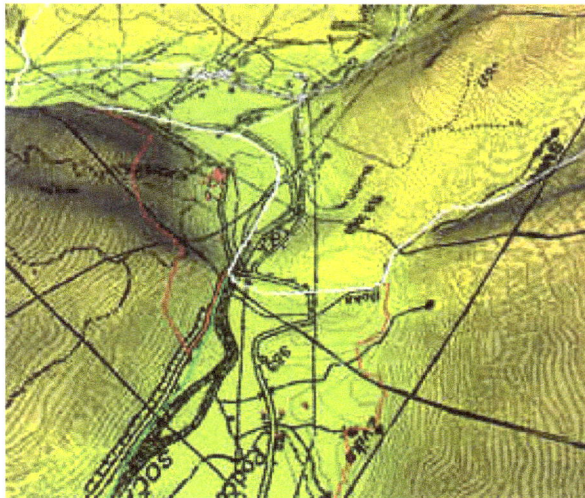
Digital Elevation Model (DEM)

Regional Geography

Regional geography is a branch of geography which studies the regions of all sizes across the Earth. It has a prevailing descriptive character. The main aim is to understand, or define the uniqueness, or character of a particular region that consists of natural as well as human elements. Attention is paid also to regionalization, which covers the proper techniques of space delimitation into regions.

Regional geography is also considered as a certain approach to study in geographical sciences

Related Fields

- Urban planning, regional planning, and spatial planning: Use the science of geography to assist in determining how to develop (or not develop) the land to meet particular criteria, such as safety, beauty, economic opportunities, the preservation of the built or natural heritage, and so on. The planning of towns, cities, and rural areas may be seen as applied geography.

- Regional science: In the 1950s, the regional science movement led by Walter Isard arose to provide a more quantitative and analytical base to geographical questions, in contrast to the descriptive tendencies of traditional geography programs. Regional science comprises the body of knowledge in which the spatial dimension plays a fundamental role, such as regional economics, resource management, location theory, urban and regional planning, transport and communication, human geography, population distribution, landscape ecology, and environmental quality.

- Interplanetary Sciences: While the discipline of geography is normally concerned with the Earth, the term can also be informally used to describe the study of other worlds, such as the planets of the Solar System and even beyond. The study of systems larger than the Earth itself usually forms part of Astronomy or Cosmology. The study of other planets is usually called planetary science. Alternative terms such as Areology (the study of Mars) have been proposed but are not widely used.

Techniques

As spatial interrelationships are key to this synoptic science, maps are a key tool. Classical cartography has been joined by a more modern approach to geographical analysis, computer-based geographic information systems (GIS).

In their study, geographers use four interrelated approaches:

- Systematic — Groups geographical knowledge into categories that can be explored globally.

- Regional — Examines systematic relationships between categories for a specific region or location on the planet.

- Descriptive — Simply specifies the locations of features and populations.

- Analytical — Asks *why we find features and populations in a specific geographic area.*

Cartography

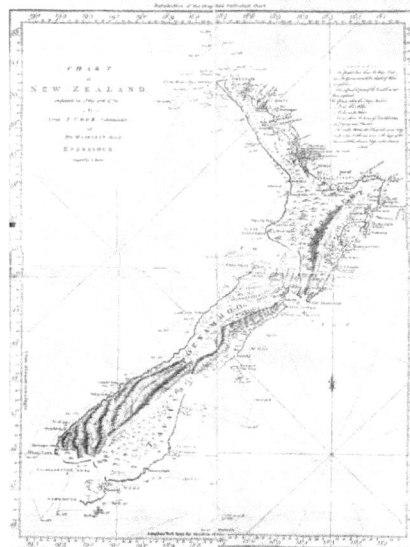

James Cook's 1770 chart of New Zealand

Cartography studies the representation of the Earth's surface with abstract symbols (map making). Although other subdisciplines of geography rely on maps for presenting their analyses, the actual making of maps is abstract enough to be regarded separately. Cartography has grown from a collection of drafting techniques into an actual science.

Cartographers must learn cognitive psychology and ergonomics to understand which symbols convey information about the Earth most effectively, and behavioural psychology to induce the readers of their maps to act on the information. They must learn geodesy and fairly advanced mathematics to understand how the shape of the Earth affects the distortion of map symbols projected onto a flat surface for viewing. It can be said, without much controversy, that cartography is the seed from which the larger field of geography grew. Most geographers will cite a childhood fascination with maps as an early sign they would end up in the field.

Geographic Information Systems

Geographic information systems (GIS) deal with the storage of information about the Earth for automatic retrieval by a computer, in an accurate manner appropriate to the information's purpose. In addition to all of the other subdisciplines of geography, GIS specialists must understand computer science and database systems. GIS has revolutionized the field of cartography: nearly all mapmaking is now done with the assistance of some form of GIS software. GIS also refers to the science of using GIS software and GIS techniques to represent, analyse, and predict the spatial relationships. In this context, GIS stands for Geographic Information Science.

Remote Sensing

Remote sensing is the science of obtaining information about Earth features from measurements made at a distance. Remotely sensed data comes in many forms, such as satellite imagery, aerial photography, and data obtained from hand-held sensors. Geographers increasingly use remotely sensed data to obtain information about the Earth's land surface, ocean, and atmosphere, because it: a) supplies objective information at a variety of spatial scales (local to global), b) provides a synoptic view of the area of interest, c) allows access to distant and inaccessible sites, d) provides spectral information outside the visible portion of the electromagnetic spectrum, and e) facilitates studies of how features/areas change over time. Remotely sensed data may be analysed either independently of, or in conjunction with other digital data layers (e.g., in a Geographic Information System).

Quantitative Methods

Geostatistics deal with quantitative data analysis, specifically the application of statistical methodology to the exploration of geographic phenomena. Geostatistics is used extensively in a variety of fields, including hydrology, geology, petroleum exploration, weather analysis, urban planning, logistics, and epidemiology. The mathematical basis for geostatistics derives from cluster analysis, linear discriminant analysis and non-parametric statistical tests, and a variety of other subjects. Applications of geostatistics rely heavily on geographic information systems, particularly for the interpolation (estimate) of unmeasured points. Geographers are making notable contributions to the method of quantitative techniques.

Qualitative Methods

Geographic qualitative methods, or ethnographical research techniques, are used by human geographers. In cultural geography there is a tradition of employing qualitative research techniques, also used in anthropology and sociology. Participant observation and in-depth interviews provide human geographers with qualitative data.

History

The oldest known world maps date back to ancient Babylon from the 9th century BC. The best known Babylonian world map, however, is the *Imago Mundi* of 600 BC. The map as reconstructed by Eckhard Unger shows Babylon on the Euphrates, surrounded by a circular landmass showing Assyria, Urartu and several cities, in turn surrounded by a "bitter river" (Oceanus), with seven islands arranged around it so as to form a seven-pointed star. The accompanying text mentions seven outer regions beyond the encircling ocean. The descriptions of five of them have survived. In contrast to the *Imago Mundi,* an earlier Babylonian world map dating back to the 9th century BC depicted Babylon as being further north from the center of the world, though it is not certain what that center was supposed to represent.

The ideas of Anaximander (c. 610 BC-c. 545 BC): considered by later Greek writers to be the true founder of geography, come to us through fragments quoted by his successors. Anaximander is credited with the invention of the gnomon, the simple, yet efficient Greek instrument that allowed the early measurement of latitude. Thales is also credited with the prediction of eclipses. The foundations of geography can be traced to the ancient cultures, such as the ancient, medieval, and early modern Chinese. The Greeks, who were the first to explore geography as both art and science, achieved this through Cartography, Philosophy, and Literature, or through Mathematics. There is some debate about who was the first person to assert that the Earth is spherical in shape, with the credit going either to Parmenides or Pythagoras. Anaxagoras was able to demonstrate that the profile of the Earth was circular by explaining eclipses. However, he still believed that the Earth was a flat disk, as did many of his contemporaries. One of the first estimates of the radius of the Earth was made by Eratosthenes.

The first rigorous system of latitude and longitude lines is credited to Hipparchus. He employed a sexagesimal system that was derived from Babylonian mathematics. The meridians were sub-divided into 360°, with each degree further subdivided 60' (minutes). To measure the longitude at different location on Earth, he suggested using eclipses to determine the relative difference in time. The extensive mapping by the Romans as they explored new lands would later provide a high level of information for Ptolemy to construct detailed atlases. He extended the work of Hipparchus, using a grid system on his maps and adopting a length of 56.5 miles for a degree.

The Ptolemy world map, reconstituted from Ptolemy's *Geographia, written c.150*

From the 3rd century onwards, Chinese methods of geographical study and writing of geographical literature became much more complex than what was found in Europe at the time (until the

13th century). Chinese geographers such as Liu An, Pei Xiu, Jia Dan, Shen Kuo, Fan Chengda, Zhou Daguan, and Xu Xiake wrote important treatises, yet by the 17th century advanced ideas and methods of Western-style geography were adopted in China.

During the Middle Ages, the fall of the Roman empire led to a shift in the evolution of geography from Europe to the Islamic world. Muslim geographers such as Muhammad al-Idrisi produced detailed world maps (such as Tabula Rogeriana), while other geographers such as Yaqut al-Hamawi, Abu Rayhan Biruni, Ibn Battuta, and Ibn Khaldun provided detailed accounts of their journeys and the geography of the regions they visited. Turkish geographer, Mahmud al-Kashgari drew a world map on a linguistic basis, and later so did Piri Reis (Piri Reis map). Further, Islamic scholars translated and interpreted the earlier works of the Romans and the Greeks and established the House of Wisdom in Baghdad for this purpose. Abū Zayd al-Balkhī, originally from Balkh, founded the "Balkhī school" of terrestrial mapping in Baghdad. Suhrāb, a late tenth century Muslim geographer accompanied a book of geographical coordinates, with instructions for making a rectangular world map with equirectangular projection or cylindrical equidistant projection.

Self portrait of Alexander von Humboldt, one of the early pioneers of geography as an academic subject in modern sense

Abu Rayhan Biruni (976-1048) first described a polar equi-azimuthal equidistant projection of the celestial sphere.[*verification needed*] He was regarded as the most skilled when it came to mapping cities and measuring the distances between them, which he did for many cities in the Middle East and the Indian subcontinent. He often combined astronomical readings and mathematical equations, in order to develop methods of pin-pointing locations by recording degrees of latitude and longitude. He also developed similar techniques when it came to measuring the heights of mountains, depths of the valleys, and expanse of the horizon. He also discussed human geography and the planetary habitability of the Earth. He also calculated the latitude of Kath, Khwarezm, using the maximum altitude of the Sun, and solved a complex geodesic equation in order to accurately compute the Earth's circumference, which were close to modern values of the Earth's circumference. His estimate of 6,339.9 km for the Earth radius was only 16.8 km less than the modern value of 6,356.7 km. In contrast to his predecessors, who measured the Earth's circumference by sighting the Sun simultaneously from two different locations, al-Biruni developed a new method of using trigonometric calculations, based on the angle between a plain and mountain top, which yielded more accurate measurements of the Earth's circumference, and made it possible for it to be measured by a single person from a single location.

The European Age of Discovery during the 16th and the 17th centuries, where many new lands were discovered and accounts by European explorers such as Christopher Columbus, Marco Polo, and James Cook revived a desire for both accurate geographic detail, and more solid theoretical foundations in Europe. The problem facing both explorers and geographers was finding the latitude and longitude of a geographic location. The problem of latitude was solved long ago but that of longitude remained; agreeing on what zero meridian should be was only part of the problem. It was left to John Harrison to solve it by inventing the chronometer H-4 in 1760, and later in 1884 for the International Meridian Conference to adopt by convention the Greenwich meridian as zero meridian.

The 18th and the 19th centuries were the times when geography became recognized as a discrete academic discipline, and became part of a typical university curriculum in Europe (especially Paris and Berlin). The development of many geographic societies also occurred during the 19th century, with the foundations of the Société de Géographie in 1821, the Royal Geographical Society in 1830, Russian Geographical Society in 1845, American Geographical Society in 1851, and the National Geographic Society in 1888. The influence of Immanuel Kant, Alexander von Humboldt, Carl Ritter, and Paul Vidal de la Blache can be seen as a major turning point in geography from a philosophy to an academic subject.

Over the past two centuries, the advancements in technology with computers have led to the development of geomatics. and new practices such as participant observation and geostatistics being incorporated into geography's portfolio of tools. In the West during the 20th century, the discipline of geography went through four major phases: environmental determinism, regional geography, the quantitative revolution, and critical geography. The strong interdisciplinary links between geography and the sciences of geology and botany, as well as economics, sociology and demographics have also grown greatly, especially as a result of Earth System Science that seeks to understand the world in a holistic view.

Notable Geographers

Gerardus Mercator

- Eratosthenes (276BC - 194BC) - calculated the size of the Earth.

- Strabo (64/63 BC – ca. AD 24) - wrote Geographica, one of the first books outlining the study of geography.

- Ptolemy (c.90–c.168) - compiled Greek and Roman knowledge into the book Geographia.

- Al Idrisi (Arabic: أبو عبد الله محمد الإدريسي; Latin: Dreses) (1100–1165/66) - author of Nuzhatul Mushtaq.

- Gerardus Mercator (1512–1594) - innovative cartographer produced the mercator projection

- Alexander von Humboldt (1769–1859) - Considered Father of modern geography, published the Kosmos and founder of the sub-field biogeography.

- Carl Ritter (1779–1859) - Considered Father of modern geography. Occupied the first chair of geography at Berlin University.

- Arnold Henry Guyot (1807–1884) - noted the structure of glaciers and advanced understanding in glacier motion, especially in fast ice flow.

- William Morris Davis (1850–1934) - father of American geography and developer of the cycle of erosion.

- Paul Vidal de la Blache (1845–1918) - founder of the French school of geopolitics and wrote the principles of human geography.

- Sir Halford John Mackinder (1861–1947) - Co-founder of the LSE, Geographical Association

- Ellen Churchill Semple (1863–1932) - She was America's first influential female geographer.

- Carl O. Sauer (1889–1975) - Prominent cultural geographer

- Walter Christaller (1893–1969) - human geographer and inventor of Central place theory.

- Yi-Fu Tuan (1930-) - Chinese-American scholar credited with starting Humanistic Geography as a discipline.

- Karl W. Butzer (1934-) - An influential German-American geographer, cultural ecologist and environmental archaeologist.

- David Harvey (1935-) - Marxist geographer and author of theories on spatial and urban geography, winner of the Vautrin Lud Prize.

- Edward Soja (1941-2015) - Noted for his work on regional development, planning and governance along with coining the terms Synekism and Postmetropolis.

- Michael Frank Goodchild (1944-) - prominent GIS scholar and winner of the RGS founder's medal in 2003.

- Doreen Massey (1944-2016) - Key scholar in the space and places of globalization and its pluralities, winner of the Vautrin Lud Prize.

- Nigel Thrift (1949-) - originator of non-representational theory.

Institutions and Societies

- American Geographical Society (U.S.)

- Anton Melik Geographical Institute (Slovenia)

- Association of American Geographers (AAG)

- National Geographic Society (U.S.)

- Royal Canadian Geographical Society (Canada)

- Royal Geographical Society (UK)

- Russian Geographical Society (Russia)

- Royal Danish Geographical Society (Denmark)

Theoretical Planetology

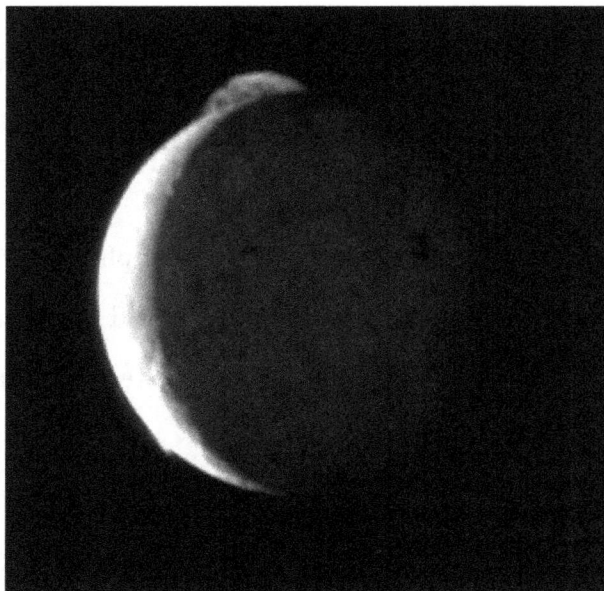

Sequence of *New Horizons* images showing Io's volcano Tvashtar spewing material 330 km above its surface. The discovery of volcanism on Io in 1979 by the Voyager 1 spacecraft confirmed the previous prediction made by theoretical planetology and is considered one of the major successes of theoretical planetology.

Theoretical planetology, also known as theoretical planetary science is a branch of planetary sciences that developed in the 20th century.

Nature of the Work

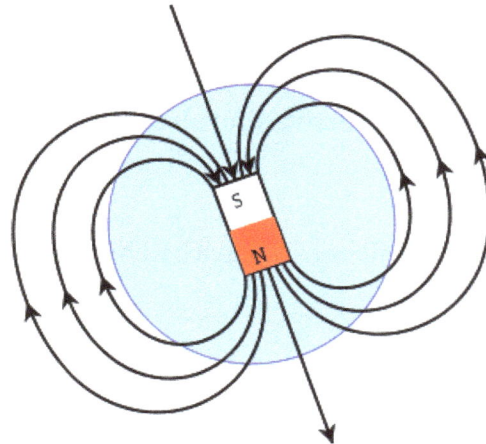

Diagram showing Earth's magnetic field: theoretical planetologists study many aspects of planetary bodies, such as how their magnetic fields are generated in their cores.

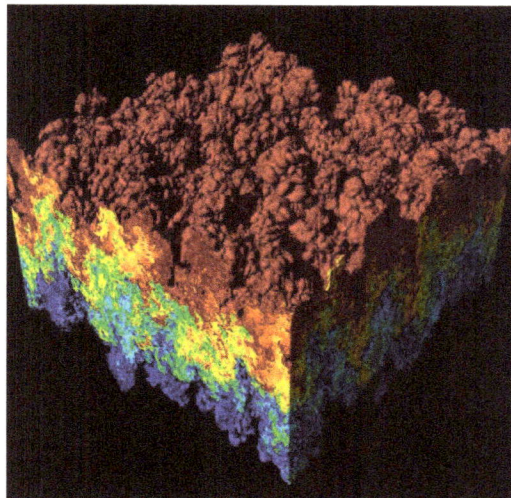

Scientific visualisation of an extremely large simulation of a Raleigh–Taylor instability caused by two mixing fluids. Theoretical planetology uses computer graphics, scientific visualisation, and fluid dynamics extensively.

Theoretical planetologists, also known as theoretical planetary scientists, use modelling techniques to develop an understanding of the internal structure of planets by making assumptions about their chemical composition and the state of their materials, then calculating the radial distribution of various properties such as temperature, pressure, or density of material across the planet's internals.

Theoretical planetologists also use numerical models to understand how the Solar System planets were formed and develop in the future, their thermal evolution, their tectonics, how magnetic fields are formed in planetary interiors, how convection processes work in the cores and mantles of terrestrial planets and in the interiors of gas giants, how their lithospheres deform, the orbital dynamics of planetary satellites, how dust and ice are transported on the surface of some planets (such as Mars), and how the atmospheric circulation takes place over a planet.

Theoretical planetologists may use laboratory experiments to understand various phenomena analogous to planetary processes, such as convection in rotating fluids.

Theoretical planetologists make extensive use of basic physics, particularly fluid dynamics and condensed matter physics, and much of their work involves interpretation of data returned by space missions, although they rarely get actively involved in them.

Educational Requirements

Typically a theoretical planetologist will have to have had higher education in physics and theoretical physics, at PhD doctorate level.

Scientific Visualisation

Because of the use of scientific visualisation animation, theoretical planetology has a relationship with computer graphics. Example movies exhibiting this relation are the 4-minute "The Origin of the Moon"

Major Successes

One of the major successes of theoretical planetology is the prediction and subsequent confirmation of volcanism on Io.

The prediction was made by Stanton Peale who wrote a scientific paper claiming that Io must be volcanically active that was published one week before Voyager 1 encountered Jupiter. When Voyager 1 photographed Io in 1979, his theory was confirmed. Later photographs of Io by the Hubble Space Telescope and from the ground also showed volcanoes on Io's surface, and they were extensively studied and photographed by the Galileo orbiter of Jupiter from 1995-2003.

Criticism

D. C. Tozer of University of Newcastle upon Tyne, writing in 1974, expressed the opinion that "it could and will be said that theoretical planetary science is a waste of time" until problems related to "sampling and scaling" are resolved, even though these problems cannot be solved by simply collecting further laboratory data.

Researchers

Researchers working on theoretical planetology include:

- David J. Stevenson

- Jonathan Lunine (University of Arizona professor of theoretical planetology and physics, and *Cassini mission scientist specialising on* Titan)

References

- Zeilik, Michael A.; Gregory, Stephan A. (1998). Introductory Astronomy & Astrophysics (4th ed.). Saunders College Publishing. p. 67. ISBN 0-03-006228-4.

- Beychok, M.R. (2005). Fundamentals Of Stack Gas Dispersion (4th ed.). author-published. ISBN 0-9644588-0-2.

- Turner, D.B. (1994). Workbook of atmospheric dispersion estimates: an introduction to dispersion modeling (2nd ed.). CRC Press. ISBN 1-56670-023-X.

- Buol, S. W.; Hole, F. D. & McCracken, R. J. (1973). Soil Genesis and Classification (First ed.). Ames, IA: Iowa State University Press. ISBN 978-0-8138-1460-5. .

- Rice, A. L. (1999). "The Challenger Expedition". Understanding the Oceans: Marine Science in the Wake of HMS Challenger. Routledge. pp. 27–48. ISBN 978-1-85728-705-9.

- Hamblin, Jacob Darwin (2005) Oceanographers and the Cold War: Disciples of Marine Science. University of Washington Press. ISBN 978-0-295-98482-7.

- Boling Guo, Daiwen Huang. Infinite-Dimensional Dynamical Systems in Atmospheric and Oceanic Science, 2014, World Scientific Publishing, ISBN 978-981-4590-37-2.

- Hughes, William. (1863). The Study of Geography. Lecture delivered at King's College, London by Sir Marc Alexander. Quoted in Baker, J.N.L (1963). The History of Geography. Oxford: Basil Blackwell. p. 66. ISBN 0-85328-022-3.

- Jean-Louis and Monique Tassoul (1920). A Concise History of Solar and Stellar Physics. London: Princeton University Press. ISBN 0-691-11711-X.

- Kurt A. Raaflaub & Richard J. A. Talbert (2009). Geography and Ethnography: Perceptions of the World in Pre-Modern Societies. John Wiley & Sons. p. 147. ISBN 1-4051-9146-5.

- "Chapter 3: Geography's Perspectives". Rediscovering Geography: New Relevance for Science and Society. Washington DC: The National Academies Press. 1997. p. 28. Retrieved 2014-05-06.

- "Sir John Murray (1841-1914) - Founder Of Modern Oceanography". Science and Engineering at The University of Edinburgh. Retrieved 7 November 2013.

- "Ocean acidification". Department of Sustainability, Environment, Water, Population & Communities: Australian Antarctic Division. 28 September 2007. Retrieved 17 April 2013.

- Hönisch, Bärbel; Ridgwell, Andy; Schmidt, Daniela N.; Thomas, E.; et al. (2012). "The Geological Record of Ocean Acidification". Science. 335 (6072): 1058–1063. Bibcode:2012Sci...335.1058H. doi:10.1126/science.1208277. PMID 22383840.

- Pielke, Roger (December 12, 2005). "Is Soil an Important Component of the Climate System?". The Climate Science Weblog. Archived from the original on 8 September 2006. Retrieved 19 April 2012.

Scientific Study of Earth and Planetary Science

Study of the major activities of the spheres that comprise the Earth's functions enable the study and understanding of spatial and temporal changes in climate, atmospheric pressure, sea-water levels and many other events. They help to correlate weather patterns as well as other instances of planetary activity and their mutual influence on each other.

Geology

Geology is an earth science comprising the study of solid Earth, the rocks of which it is composed, and the processes by which they change. Geology can also refer generally to the study of the solid features of any celestial body (such as the geology of the Moon or Mars).

Geology gives insight into the history of the Earth by providing the primary evidence for plate tectonics, the evolutionary history of life, and past climates. Geology is important for mineral and hydrocarbon exploration and exploitation, evaluating water resources, understanding of natural hazards, the remediation of environmental problems, and for providing insights into past climate change. Geology also plays a role in geotechnical engineering and is a major academic discipline.

Geologic Materials

The majority of geological data comes from research on solid Earth materials. These typically fall into one of two categories: rock and unconsolidated material.

There are three major types of rock: igneous, sedimentary, and metamorphic. The rock cycle is an important concept in geology which illustrates the relationships between these three types of rock, and magma. When a rock crystallizes from melt (magma and/or lava), it is an igneous rock. This rock can be weathered and eroded, and then redeposited and lithified into a sedimentary rock, or be turned into a metamorphic rock due to heat and pressure that change the mineral content of the rock which gives it a characteristic fabric. The sedimentary rock can then be subsequently turned into a metamorphic rock due to heat and pressure and is then weathered, eroded, deposited, and lithified, ultimately becoming a sedimentary rock. Sedimentary rock may also be re-eroded and redeposited, and metamorphic rock may also undergo additional metamorphism. All three types of rocks may be re-melted; when this happens, a new magma is formed, from which an igneous rock may once again crystallize.

Rock

Rock Cycle

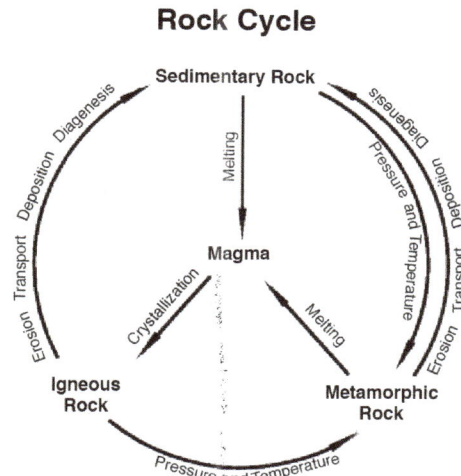

This schematic diagram of the rock cycle shows the relationship between magma and sedimentary, metamorphic, and igneous rock

The majority of research in geology is associated with the study of rock, as rock provides the primary record of the majority of the geologic history of the Earth.

Unconsolidated Material

Geologists also study unlithified material, which typically comes from more recent deposits. These materials are superficial deposits which lie above the bedrock. Because of this, the study of such material is often known as Quaternary geology, after the recent Quaternary Period. This includes the study of sediment and soils, including studies in geomorphology, sedimentology, and paleoclimatology.

Whole-Earth Structure

Plate Tectonics

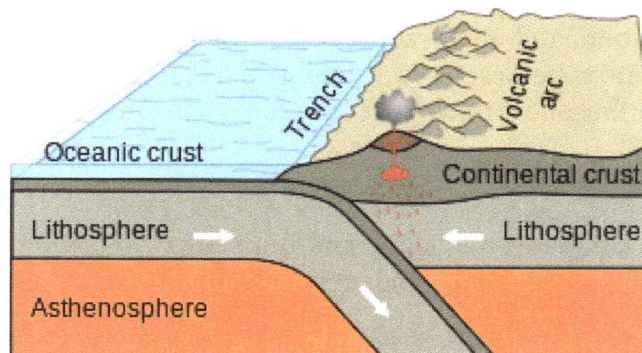

Oceanic-continental convergence resulting in subduction and volcanic arcs illustrates one effect of plate tectonics.

In the 1960s, a series of discoveries, the most important of which was seafloor spreading, showed that the Earth's lithosphere, which includes the crust and rigid uppermost portion of the upper mantle, is separated into a number of tectonic plates that move across the plastically deforming, solid, upper mantle, which is called the asthenosphere. There is an intimate coupling between the movement of the plates on the surface and the convection of the mantle: oceanic plate motions and mantle convection currents always move in the same direction, because the oceanic lithosphere is the rigid upper thermal boundary layer of the convecting mantle. This coupling between rigid plates moving on the surface of the Earth and the convecting mantle is called plate tectonics.

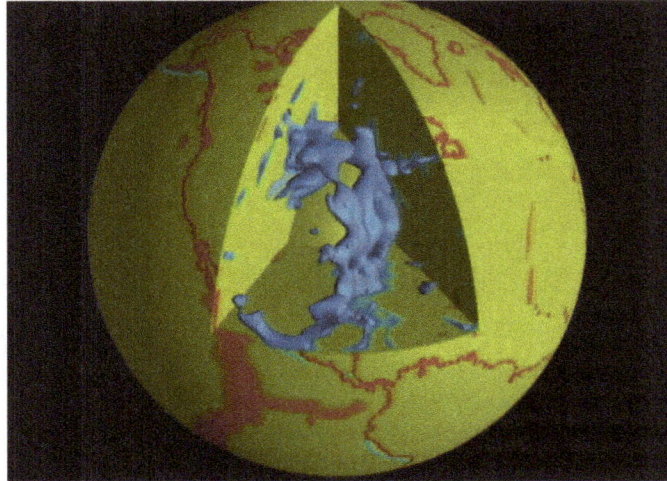

On this diagram, subducting slabs are in blue, and continental margins and a few plate boundaries are in red. The blue blob in the cutaway section is the seismically imaged Farallon Plate, which is subducting beneath North America. The remnants of this plate on the Surface of the Earth are the Juan de Fuca Plate and Explorer plate in the Northwestern USA / Southwestern Canada, and the Cocos Plate on the west coast of Mexico.

The development of plate tectonics provided a physical basis for many observations of the solid Earth. Long linear regions of geologic features could be explained as plate boundaries. Mid-ocean ridges, high regions on the seafloor where hydrothermal vents and volcanoes exist, were explained as divergent boundaries, where two plates move apart. Arcs of volcanoes and earthquakes were explained as convergent boundaries, where one plate subducts under another. Transform boundaries, such as the San Andreas Fault system, resulted in widespread powerful earthquakes. Plate tectonics also provided a mechanism for Alfred Wegener's theory of continental drift, in which the continents move across the surface of the Earth over geologic time. They also provided a driving force for crustal deformation, and a new setting for the observations of structural geology. The power of the theory of plate tectonics lies in its ability to combine all of these observations into a single theory of how the lithosphere moves over the convecting mantle.

Earth Structure

Advances in seismology, computer modeling, and mineralogy and crystallography at high temperatures and pressures give insights into the internal composition and structure of the Earth.

Seismologists can use the arrival times of seismic waves in reverse to image the interior of the Earth. Early advances in this field showed the existence of a liquid outer core (where shear waves were not able to propagate) and a dense solid inner core. These advances led to the development of a layered model of the Earth, with a crust and lithosphere on top, the mantle below (separated within itself by seismic discontinuities at 410 and 660 kilometers), and the outer core and inner core below that. More recently, seismologists have been able to create detailed images of wave speeds inside the earth in the same way a doctor images a body in a CT scan. These images have led to a much more detailed view of the interior of the Earth, and have replaced the simplified layered model with a much more dynamic model.

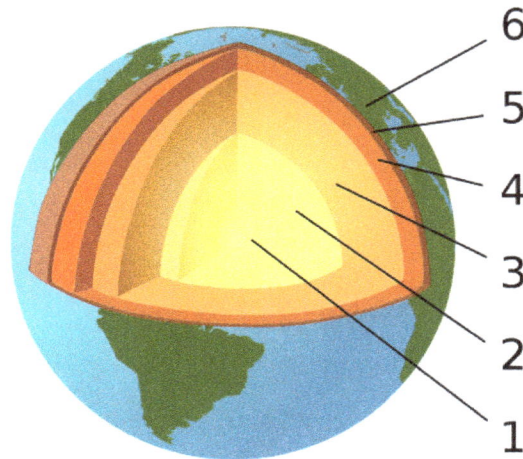

The Earth's layered structure. (1) inner core; (2) outer core; (3) lower mantle; (4) upper mantle; (5) lithosphere; (6) crust (part of the lithosphere)

Mineralogists have been able to use the pressure and temperature data from the seismic and modelling studies alongside knowledge of the elemental composition of the Earth to reproduce these conditions in experimental settings and measure changes in crystal structure. These studies explain the chemical changes associated with the major seismic discontinuities in the mantle and show the crystallographic structures expected in the inner core of the Earth.

Geologic Time

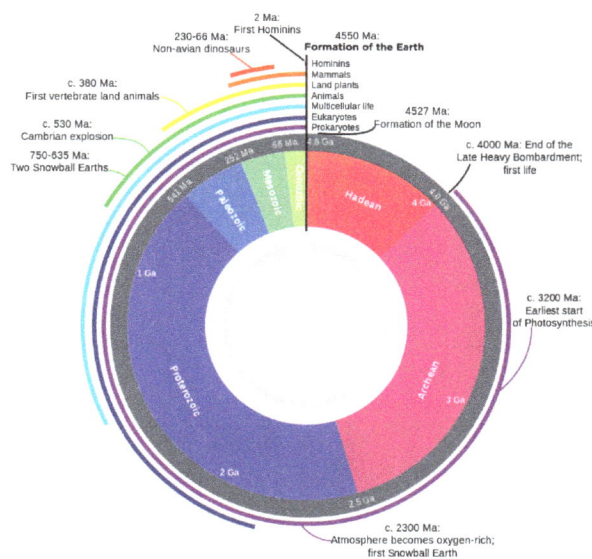

Geological time put in a diagram called a geological clock, showing the relative lengths of the eons of the Earth's history.

The geologic time scale encompasses the history of the Earth. It is bracketed at the old end by the dates of the earliest Solar System material at 4.567 Ga, (gigaannum: billion years ago) and the age of the Earth at 4.54 Ga at the beginning of the informally recognized Hadean eon. At the young end of the scale, it is bracketed by the present day in the Holocene epoch.

Important Milestones

- 4.567 Ga: Solar system formation

- 4.54 Ga: Accretion of Earth

- c. 4 Ga: End of Late Heavy Bombardment, first life

- c. 3.5 Ga: Start of photosynthesis

- c. 2.3 Ga: Oxygenated atmosphere, first snowball Earth

- 730–635 Ma (megaannum: million years ago): second snowball Earth

- 542 ± 0.3 Ma: Cambrian explosion – vast multiplication of hard-bodied life; first abundant fossils; start of the Paleozoic

- c. 380 Ma: First vertebrate land animals

- 250 Ma: Permian-Triassic extinction – 90% of all land animals die; end of Paleozoic and beginning of Mesozoic

- 66 Ma: Cretaceous–Paleogene extinction – Dinosaurs die; end of Mesozoic and beginning of Cenozoic

- c. 7 Ma: First hominins appear

- 3.9 Ma: First Australopithecus, direct ancestor to modern Homo sapiens, appear

- 200 ka (kiloannum: thousand years ago): First modern Homo sapiens appear in East Africa

Brief Time Scale

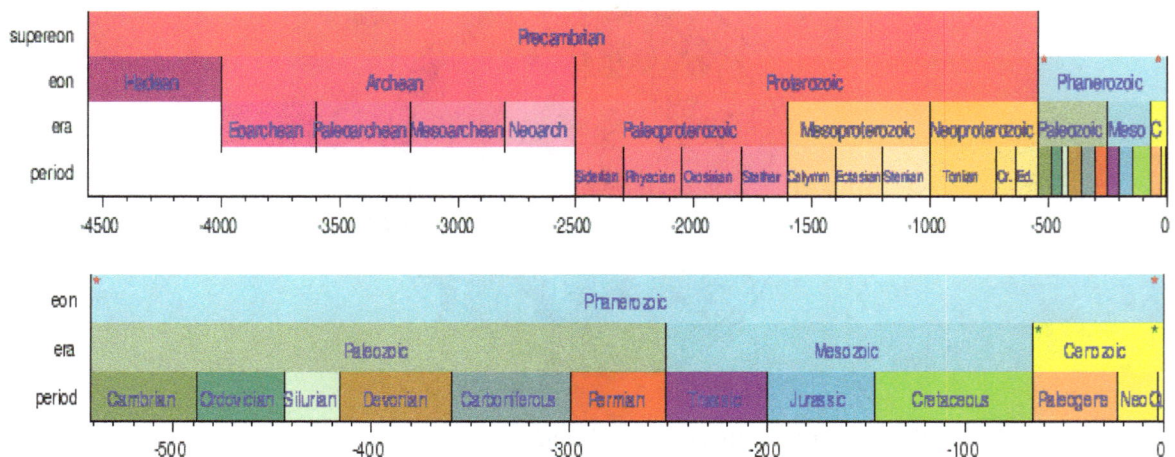

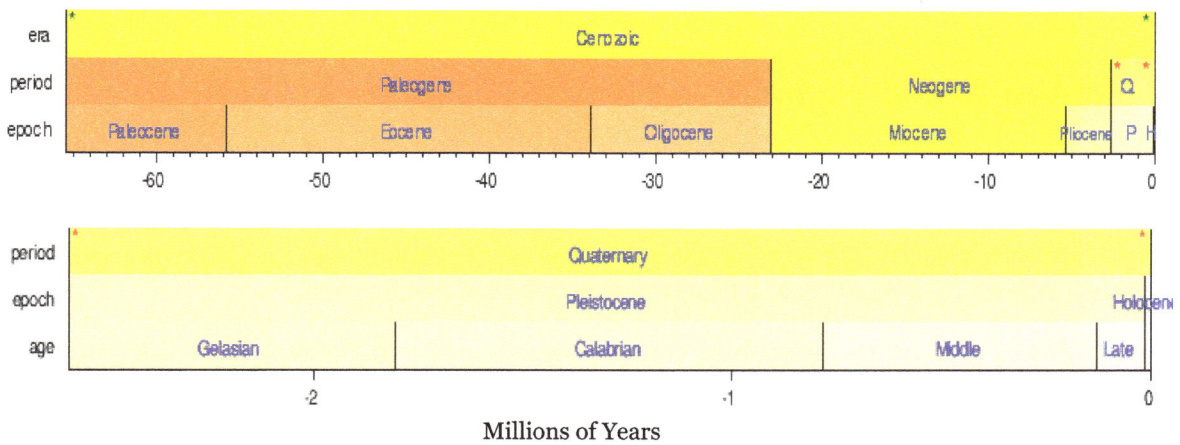

The following four timelines show the geologic time scale. The first shows the entire time from the formation of the Earth to the present, but this compresses the most recent eon. Therefore, the second scale shows the most recent eon with an expanded scale. The second scale compresses the most recent era, so the most recent era is expanded in the third scale. Since the Quaternary is a very short period with short epochs, it is further expanded in the fourth scale. The second, third, and fourth timelines are therefore each subsections of their preceding timeline as indicated by asterisks. The Holocene (the latest epoch) is too small to be shown clearly on the third timeline on the right, another reason for expanding the fourth scale. The Pleistocene (P) epoch. Q stands for the Quaternary period.

Dating Methods

Geologists use a variety of methods to give both relative and absolute dates to geological events. They then use these dates to find the rates at which processes occur.

Relative Dating

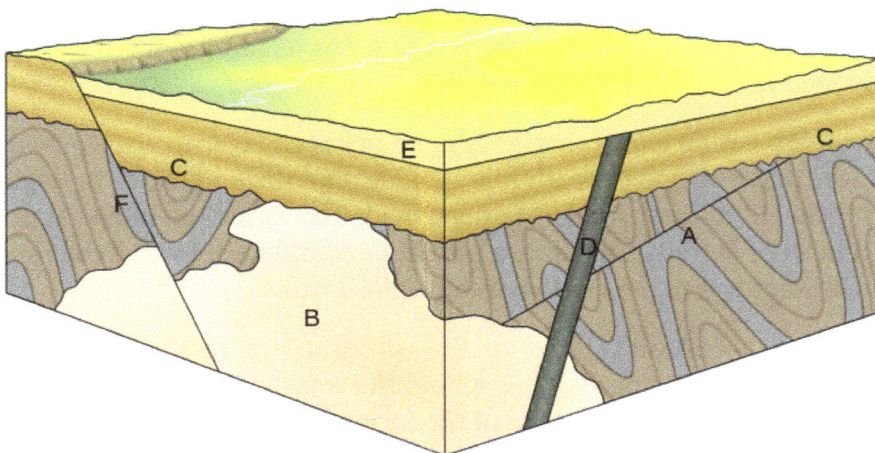

Cross-cutting relations can be used to determine the relative ages of rock strata and other geological structures. Explanations: A – folded rock strata cut by a thrust fault; B – large intrusion (cutting through A); C – erosional angular unconformity (cutting off A & B) on which rock strata were deposited; D – volcanic dyke (cutting through A, B & C); E – even younger rock strata (overlying C & D); F – normal fault (cutting through A, B, C & E).

Methods for relative dating were developed when geology first emerged as a formal science. Geologists still use the following principles today as a means to provide information about geologic history and the timing of geologic events.

The principle of Uniformitarianism states that the geologic processes observed in operation that modify the Earth's crust at present have worked in much the same way over geologic time. A fundamental principle of geology advanced by the 18th century Scottish physician and geologist James Hutton, is that "the present is the key to the past." In Hutton's words: "the past history of our globe must be explained by what can be seen to be happening now."

The principle of intrusive relationships concerns crosscutting intrusions. In geology, when an igneous intrusion cuts across a formation of sedimentary rock, it can be determined that the igneous intrusion is younger than the sedimentary rock. There are a number of different types of intrusions, including stocks, laccoliths, batholiths, sills and dikes.

The principle of cross-cutting relationships pertains to the formation of faults and the age of the sequences through which they cut. Faults are younger than the rocks they cut; accordingly, if a fault is found that penetrates some formations but not those on top of it, then the formations that were cut are older than the fault, and the ones that are not cut must be younger than the fault. Finding the key bed in these situations may help determine whether the fault is a normal fault or a thrust fault.

The principle of inclusions and components states that, with sedimentary rocks, if inclusions (or clasts) are found in a formation, then the inclusions must be older than the formation that contains them. For example, in sedimentary rocks, it is common for gravel from an older formation to be ripped up and included in a newer layer. A similar situation with igneous rocks occurs when xenoliths are found. These foreign bodies are picked up as magma or lava flows, and are incorporated, later to cool in the matrix. As a result, xenoliths are older than the rock which contains them.

The principle of original horizontality states that the deposition of sediments occurs as essentially horizontal beds. Observation of modern marine and non-marine sediments in a wide variety of environments supports this generalization (although cross-bedding is inclined, the overall orientation of cross-bedded units is horizontal).

The principle of superposition states that a sedimentary rock layer in a tectonically undisturbed sequence is younger than the one beneath it and older than the one above it. Logically a younger layer cannot slip beneath a layer previously deposited. This principle allows sedimentary layers to be viewed as a form of vertical time line, a partial or complete record of the time elapsed from deposition of the lowest layer to deposition of the highest bed.

The principle of faunal succession is based on the appearance of fossils in sedimentary rocks. As organisms exist at the same time period throughout the world, their presence or (sometimes) absence may be used to provide a relative age of the formations in which they are found. Based on principles laid out by William Smith almost a hundred years before the publication of Charles Darwin's theory of evolution, the principles of succession were developed independently of evolutionary thought. The principle becomes quite complex, however, given the uncertainties of fossilization, the localization of fossil types due to lateral changes in habitat (facies change in sedimentary strata), and that not all fossils may be found globally at the same time.

Absolute Dating

Geologists also use methods to determine the absolute age of rock samples and geological events. These dates are useful on their own and may also be used in conjunction with relative dating methods or to calibrate relative methods.

At the beginning of the 20th century, important advancement in geological science was facilitated by the ability to obtain accurate absolute dates to geologic events using radioactive isotopes and other methods. This changed the understanding of geologic time. Previously, geologists could only use fossils and stratigraphic correlation to date sections of rock relative to one another. With isotopic dates it became possible to assign absolute ages to rock units, and these absolute dates could be applied to fossil sequences in which there was datable material, converting the old relative ages into new absolute ages.

For many geologic applications, isotope ratios of radioactive elements are measured in minerals that give the amount of time that has passed since a rock passed through its particular closure temperature, the point at which different radiometric isotopes stop diffusing into and out of the crystal lattice. These are used in geochronologic and thermochronologic studies. Common methods include uranium-lead dating, potassium-argon dating, argon-argon dating and uranium-thorium dating. These methods are used for a variety of applications. Dating of lava and volcanic ash layers found within a stratigraphic sequence can provide absolute age data for sedimentary rock units which do not contain radioactive isotopes and calibrate relative dating techniques. These methods can also be used to determine ages of pluton emplacement. Thermochemical techniques can be used to determine temperature profiles within the crust, the uplift of mountain ranges, and paleotopography.

Fractionation of the lanthanide series elements is used to compute ages since rocks were removed from the mantle.

Other methods are used for more recent events. Optically stimulated luminescence and cosmogenic radionucleide dating are used to date surfaces and/or erosion rates. Dendrochronology can also be used for the dating of landscapes. Radiocarbon dating is used for geologically young materials containing organic carbon.

Geological Development of an Area

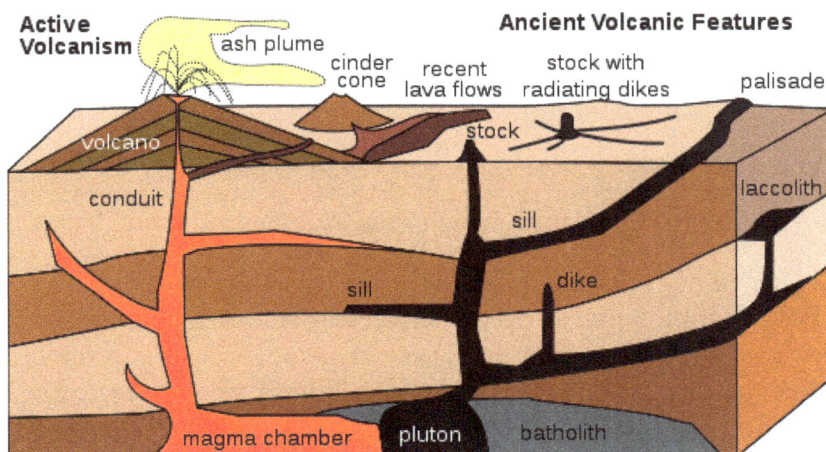

An originally horizontal sequence of sedimentary rocks (in shades of tan) are affected by igneous activity. Deep below the surface are a magma chamber and large associated igneous bodies. The magma chamber feeds the volcano, and sends offshoots of magma that will later crystallize into dikes and sills. Magma also advances upwards to form intrusive igneous bodies. The diagram illustrates both a cinder cone volcano, which releases ash, and a composite volcano, which releases both lava and ash.

The geology of an area changes through time as rock units are deposited and inserted and deformational processes change their shapes and locations.

Rock units are first emplaced either by deposition onto the surface or intrusion into the overlying rock. Deposition can occur when sediments settle onto the surface of the Earth and later lithify into sedimentary rock, or when as volcanic material such as volcanic ash or lava flows blanket the surface. Igneous intrusions such as batholiths, laccoliths, dikes, and sills, push upwards into the overlying rock, and crystallize as they intrude.

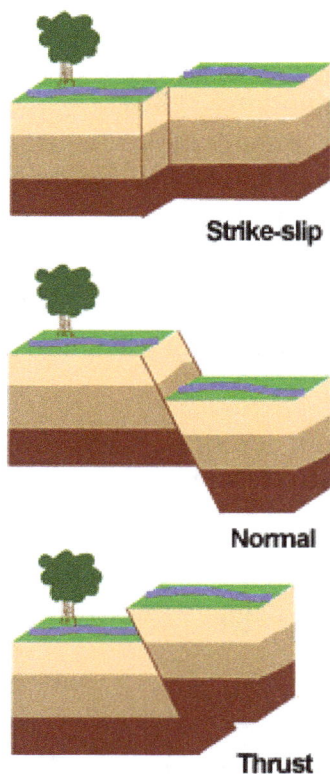

Strike-slip

Normal

Thrust

An illustration of the three types of faults. Strike-slip faults occur when rock units slide past one another, normal faults occur when rocks are undergoing horizontal extension, and thrust faults occur when rocks are undergoing horizontal shortening.

After the initial sequence of rocks has been deposited, the rock units can be deformed and/or metamorphosed. Deformation typically occurs as a result of horizontal shortening, horizontal extension, or side-to-side (strike-slip) motion. These structural regimes broadly relate to convergent boundaries, divergent boundaries, and transform boundaries, respectively, between tectonic plates.

When rock units are placed under horizontal compression, they shorten and become thicker. Because rock units, other than muds, do not significantly change in volume, this is accomplished in two primary ways: through faulting and folding. In the shallow crust, where brittle deformation can occur, thrust faults form, which cause deeper rock to move on top of shallower rock. Because deeper rock is often older, as noted by the principle of superposition, this can result in older rocks moving on top of younger ones. Movement along faults can result in folding, either because the faults are not planar or because rock layers are dragged along, forming drag folds as slip occurs along the fault. Deeper in the Earth, rocks behave plastically, and fold instead of faulting. These folds can either be those where the material in the center of the fold buckles upwards, creating "antiforms", or where it buckles downwards, creating "synforms". If the tops of the rock units within the folds remain pointing upwards, they are called anticlines and synclines, respectively. If some of the units in the fold are facing downward, the structure is called an overturned anticline or syncline, and if all of the rock units are overturned or the correct up-direction is unknown, they are simply called by the most general terms, antiforms and synforms.

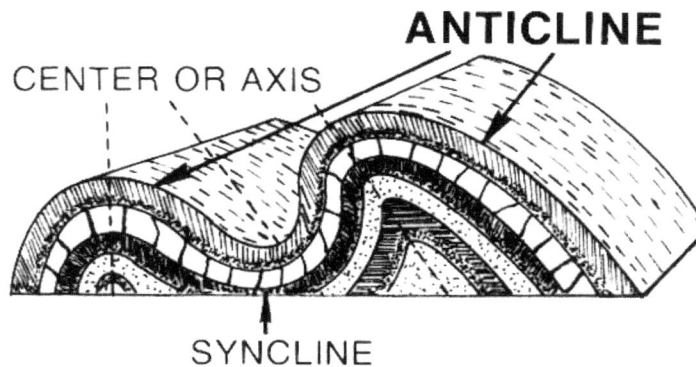

A diagram of folds, indicating an anticline and a syncline.

Even higher pressures and temperatures during horizontal shortening can cause both folding and metamorphism of the rocks. This metamorphism causes changes in the mineral composition of the rocks; creates a foliation, or planar surface, that is related to mineral growth under stress. This can remove signs of the original textures of the rocks, such as bedding in sedimentary rocks, flow features of lavas, and crystal patterns in crystalline rocks.

Extension causes the rock units as a whole to become longer and thinner. This is primarily accomplished through normal faulting and through the ductile stretching and thinning. Normal faults drop rock units that are higher below those that are lower. This typically results in younger units being placed below older units. Stretching of units can result in their thinning; in fact, there is a location within the Maria Fold and Thrust Belt in which the entire sedimentary sequence of the Grand Canyon can be seen over a length of less than a meter. Rocks at the depth to be ductilely stretched are often also metamorphosed. These stretched rocks can also pinch into lenses, known as boudins, after the French word for "sausage", because of their visual similarity.

Where rock units slide past one another, strike-slip faults develop in shallow regions, and become shear zones at deeper depths where the rocks deform ductilely.

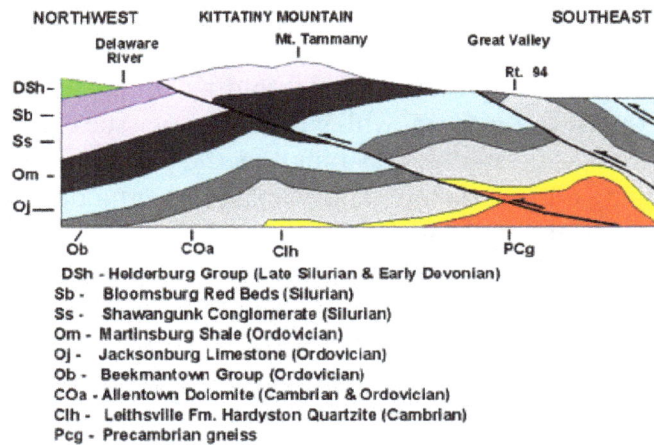

Geologic cross section of Kittatinny Mountain. This cross section shows metamorphic rocks, overlain
by younger sediments deposited after the metamorphic event. These rock units were later
folded and faulted during the uplift of the mountain.

The addition of new rock units, both depositionally and intrusively, often occurs during deformation. Faulting and other deformational processes result in the creation of topographic gradients, causing material on the rock unit that is increasing in elevation to be eroded by hillslopes and channels. These sediments are deposited on the rock unit that is going down. Continual motion along the fault maintains the topographic gradient in spite of the movement of sediment, and continues to create accommodation space for the material to deposit. Deformational events are often also associated with volcanism and igneous activity. Volcanic ashes and lavas accumulate on the surface, and igneous intrusions enter from below. Dikes, long, planar igneous intrusions, enter along cracks, and therefore often form in large numbers in areas that are being actively deformed. This can result in the emplacement of dike swarms, such as those that are observable across the Canadian shield, or rings of dikes around the lava tube of a volcano.

All of these processes do not necessarily occur in a single environment, and do not necessarily occur in a single order. The Hawaiian Islands, for example, consist almost entirely of layered basaltic lava flows. The sedimentary sequences of the mid-continental United States and the Grand Canyon in the southwestern United States contain almost-undeformed stacks of sedimentary rocks that have remained in place since Cambrian time. Other areas are much more geologically complex. In the southwestern United States, sedimentary, volcanic, and intrusive rocks have been metamorphosed, faulted, foliated, and folded. Even older rocks, such as the Acasta gneiss of the Slave craton in northwestern Canada, the oldest known rock in the world have been metamorphosed to the point where their origin is undiscernable without laboratory analysis. In addition, these processes can occur in stages. In many places, the Grand Canyon in the southwestern United States being a very visible example, the lower rock units were metamorphosed and deformed, and then deformation ended and the upper, undeformed units were deposited. Although any amount of rock emplacement and rock deformation can occur, and they can occur any number of times, these concepts provide a guide to understanding the geological history of an area.

Methods of Geology

Geologists use a number of field, laboratory, and numerical modeling methods to decipher Earth history and understand the processes that occur on and inside the Earth. In typical geological investigations, geologists use primary information related to petrology (the study of rocks), stratigraphy (the study of sedimentary layers), and structural geology (the study of positions of rock units and their deformation). In many cases, geologists also study modern soils, rivers, landscapes, and glaciers; investigate past and current life and biogeochemical pathways, and use geophysical methods to investigate the subsurface.

Field Methods

A standard Brunton Pocket Transit, commonly used by geologists for mapping and surveying.

A typical USGS field mapping camp in the 1950s

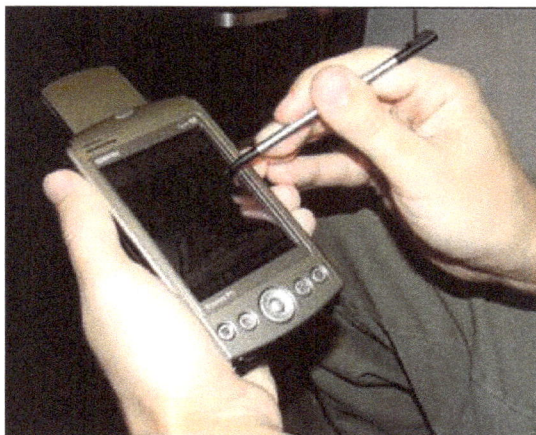

Today, handheld computers with GPS and geographic information systems software are often used in geological field work (digital geologic mapping).

Geological field work varies depending on the task at hand. Typical fieldwork could consist of:

- Geological mapping

    o Structural mapping: the locations of major rock units and the faults and folds that led to their placement there.

    o Stratigraphic mapping: the locations of sedimentary facies (lithofacies and biofacies) or the mapping of isopachs of equal thickness of sedimentary rock

    o Surficial mapping: the locations of soils and surficial deposits

- Surveying of topographic features

    o Creation of topographic maps

    o Work to understand change across landscapes, including:

        ▪ Patterns of erosion and deposition

        ▪ River channel change through migration and avulsion

        ▪ Hillslope processes

- Subsurface mapping through geophysical methods

    o These methods include:

        ▪ Shallow seismic surveys

        ▪ Ground-penetrating radar

        ▪ Aeromagnetic surveys

        ▪ Electrical resistivity tomography

    o They are used for:

        ▪ Hydrocarbon exploration

        ▪ Finding groundwater

        ▪ Locating buried archaeological artifacts

- High-resolution stratigraphy

    o Measuring and describing stratigraphic sections on the surface

    o Well drilling and logging

- Biogeochemistry and geomicrobiology

    o Collecting samples to:

        ▪ Determine biochemical pathways

- Identify new species of organisms

- Identify new chemical compounds

  o And to use these discoveries to:

    - Understand early life on Earth and how it functioned and metabolized

    - Find important compounds for use in pharmaceuticals.

- Paleontology: excavation of fossil material

  o For research into past life and evolution

  o For museums and education

- Collection of samples for geochronology and thermochronology

- Glaciology: measurement of characteristics of glaciers and their motion

Petrology

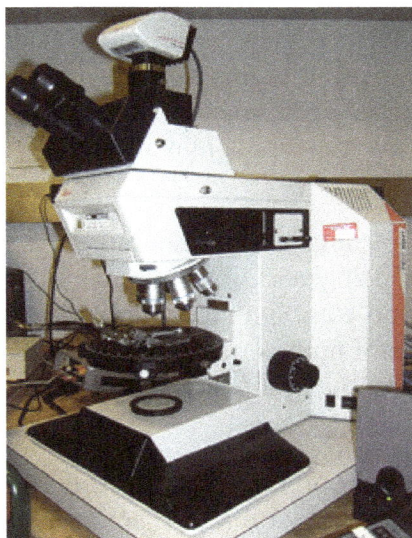

A petrographic microscope, which is an optical microscope fitted with cross-polarizing lenses, a conoscopic lens, and compensators (plates of anisotropic materials; gypsum plates and quartz wedges are common), for crystallographic analysis.

In addition to identifying rocks in the field, petrologists identify rock samples in the laboratory. Two of the primary methods for identifying rocks in the laboratory are through optical microscopy and by using an electron microprobe. In an optical mineralogy analysis, thin sections of rock samples are analyzed through a petrographic microscope, where the minerals can be identified through their different properties in plane-polarized and cross-polarized light, including their birefringence, pleochroism, twinning, and interference properties with a conoscopic lens. In the electron microprobe, individual locations are analyzed for their exact chemical compositions and variation in composition within individual crystals. Stable and radioactive isotope studies provide insight into the geochemical evolution of rock units.

Petrologists can also use fluid inclusion data and perform high temperature and pressure physical experiments to understand the temperatures and pressures at which different mineral phases appear, and how they change through igneous and metamorphic processes. This research can be extrapolated to the field to understand metamorphic processes and the conditions of crystallization of igneous rocks. This work can also help to explain processes that occur within the Earth, such as subduction and magma chamber evolution.

Structural Geology

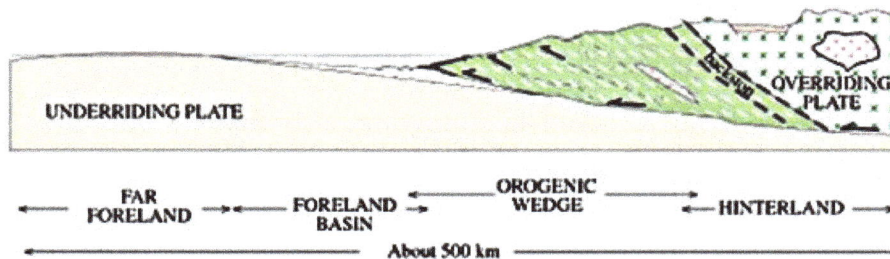

A diagram of an orogenic wedge. The wedge grows through faulting in the interior and along the main basal fault, called the décollement. It builds its shape into a critical taper, in which the angles within the wedge remain the same as failures inside the material balance failures along the décollement. It is analogous to a bulldozer pushing a pile of dirt, where the bulldozer is the overriding plate.

Structural geologists use microscopic analysis of oriented thin sections of geologic samples to observe the fabric within the rocks which gives information about strain within the crystalline structure of the rocks. They also plot and combine measurements of geological structures in order to better understand the orientations of faults and folds in order to reconstruct the history of rock deformation in the area. In addition, they perform analog and numerical experiments of rock deformation in large and small settings.

The analysis of structures is often accomplished by plotting the orientations of various features onto stereonets. A stereonet is a stereographic projection of a sphere onto a plane, in which planes are projected as lines and lines are projected as points. These can be used to find the locations of fold axes, relationships between faults, and relationships between other geologic structures.

Among the most well-known experiments in structural geology are those involving orogenic wedges, which are zones in which mountains are built along convergent tectonic plate boundaries. In the analog versions of these experiments, horizontal layers of sand are pulled along a lower surface into a back stop, which results in realistic-looking patterns of faulting and the growth of a critically tapered (all angles remain the same) orogenic wedge. Numerical models work in the same way as these analog models, though they are often more sophisticated and can include patterns of erosion and uplift in the mountain belt. This helps to show the relationship between erosion and the shape of the mountain range. These studies can also give useful information about pathways for metamorphism through pressure, temperature, space, and time.

Stratigraphy

In the laboratory, stratigraphers analyze samples of stratigraphic sections that can be returned from the field, such as those from drill cores. Stratigraphers also analyze data from geophysical surveys that show the locations of stratigraphic units in the subsurface. Geophysical data and well logs can be combined to produce a better view of the subsurface, and stratigraphers often use computer programs to do this in three dimensions. Stratigraphers can then use these data to reconstruct ancient processes occurring on the surface of the Earth, interpret past environments, and locate areas for water, coal, and hydrocarbon extraction.

In the laboratory, biostratigraphers analyze rock samples from outcrop and drill cores for the fossils found in them. These fossils help scientists to date the core and to understand the depositional environment in which the rock units formed. Geochronologists precisely date rocks within the stratigraphic section in order to provide better absolute bounds on the timing and rates of deposition. Magnetic stratigraphers look for signs of magnetic reversals in igneous rock units within the drill cores. Other scientists perform stable isotope studies on the rocks to gain information about past climate.

Planetary Geology

Surface of Mars as photographed by the Viking 2 lander December 9, 1977.

With the advent of space exploration in the twentieth century, geologists have begun to look at other planetary bodies in the same ways that have been developed to study the Earth. This new field of study is called planetary geology (sometimes known as astrogeology) and relies on known geologic principles to study other bodies of the solar system.

Although the Greek-language-origin prefix geo refers to Earth, "geology" is often used in conjunction with the names of other planetary bodies when describing their composition and internal processes: examples are "the geology of Mars" and "Lunar geology". Specialised terms such as selenology (studies of the Moon), areology (of Mars), etc., are also in use.

Although planetary geologists are interested in studying all aspects of other planets, a significant focus is to search for evidence of past or present life on other worlds. This has led to many missions whose primary or ancillary purpose is to examine planetary bodies for evidence of life. One of these is the Phoenix lander, which analyzed Martian polar soil for water, chemical, and mineralogical constituents related to biological processes.

Applied Geology

Economic Geology

Economic geology is an important branch of geology which deals with different aspects of economic minerals being used by humankind to fulfill its various needs. The economic minerals are those which can be extracted profitably. Economic geologists help locate and manage the Earth's natural resources, such as petroleum and coal, as well as mineral resources, which include metals such as iron, copper, and uranium.

Mining Geology

Mining geology consists of the extractions of mineral resources from the Earth. Some resources of economic interests include gemstones, metals, and many minerals such as asbestos, perlite, mica, phosphates, zeolites, clay, pumice, quartz, and silica, as well as elements such as sulfur, chlorine, and helium.

Petroleum Geology

Mud log in process, a common way to study the lithology when drilling oil wells.

Petroleum geologists study the locations of the subsurface of the Earth which can contain extractable hydrocarbons, especially petroleum and natural gas. Because many of these reservoirs are found in sedimentary basins, they study the formation of these basins, as well as their sedimentary and tectonic evolution and the present-day positions of the rock units.

Engineering Geology

Engineering geology is the application of the geologic principles to engineering practice for the purpose of assuring that the geologic factors affecting the location, design, construction, operation, and maintenance of engineering works are properly addressed.

In the field of civil engineering, geological principles and analyses are used in order to ascertain the mechanical principles of the material on which structures are built. This allows tunnels to be built without collapsing, bridges and skyscrapers to be built with sturdy foundations, and buildings to be built that will not settle in clay and mud.

Hydrology and Environmental Issues

Geology and geologic principles can be applied to various environmental problems such as stream restoration, the restoration of brownfields, and the understanding of the interaction between natural habitat and the geologic environment. Groundwater hydrology, or hydrogeology, is used to locate groundwater, which can often provide a ready supply of uncontaminated water and is especially important in arid regions, and to monitor the spread of contaminants in groundwater wells.

Geologists also obtain data through stratigraphy, boreholes, core samples, and ice cores. Ice cores and sediment cores are used to for paleoclimate reconstructions, which tell geologists about past and present temperature, precipitation, and sea level across the globe. These datasets are our primary source of information on global climate change outside of instrumental data.

Natural Hazards

Geologists and geophysicists study natural hazards in order to enact safe building codes and warning systems that are used to prevent loss of property and life. Examples of important natural hazards that are pertinent to geology (as opposed those that are mainly or only pertinent to meteorology) are:

Rockfall in the Grand Canyon

- Avalanches

- Earthquakes

- Floods

- Landslides and debris flows

- River channel migration and avulsion

- Liquefaction

- Sinkholes

- Subsidence

- Tsunamis

- Volcanoes

History of Geology

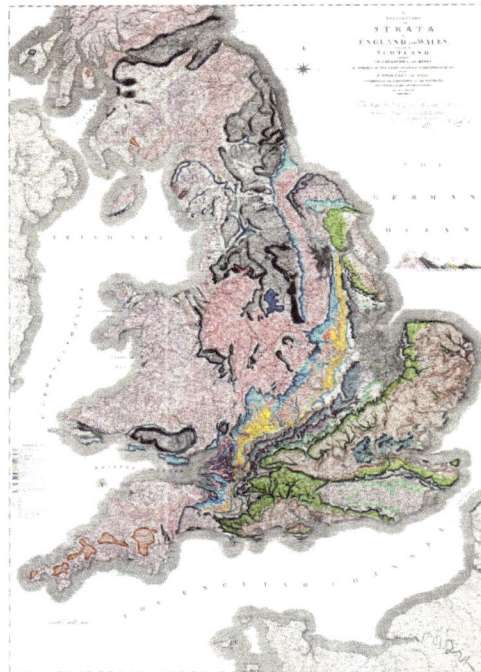

William Smith's geologic map of England, Wales, and southern Scotland. Completed in 1815, it was the second national-scale geologic map, and by far the most accurate of its time.

The study of the physical material of the Earth dates back at least to ancient Greece when Theophrastus (372–287 BCE) wrote the work *Peri Lithon (On Stones). During the* Roman period, Pliny the Elder wrote in detail of the many minerals and metals then in practical use – even correctly noting the origin of amber.

Some modern scholars, such as Fielding H. Garrison, are of the opinion that the origin of the science of geology can be traced to Persia after the Muslim conquests had come to an end. Abu al-Rayhan al-Biruni (973–1048 CE) was one of the earliest Persian geologists, whose works included the earliest writings on the geology of India, hypothesizing that the Indian subcontinent was once a sea. Drawing from Greek and Indian scientific literature that were not destroyed by the Muslim conquests, the Persian scholar Ibn Sina (Avicenna, 981–1037) proposed detailed explanations for the formation of mountains, the origin of earthquakes, and other topics central to modern geology, which provided an essential foundation for the later development of the science. In China, the polymath Shen Kuo (1031–1095) formulated a hypothesis for the process of land formation: based on his observation of fossil animal shells in a geological stratum in a mountain hundreds of miles from the ocean, he inferred that the land was formed by erosion of the mountains and by deposition of silt.

Nicolas Steno (1638–1686) is credited with the law of superposition, the principle of original horizontality, and the principle of lateral continuity: three defining principles of stratigraphy.

The word *geology* was first used by Ulisse Aldrovandi in 1603, then by Jean-André Deluc in 1778 and introduced as a fixed term by Horace-Bénédict de Saussure in 1779. But according to another source, the word "geology" comes from a Norwegian, Mikkel Pedersøn Escholt (1600–1699), who was a priest and scholar. Escholt first used the definition in his book titled, *Geologica Norvegica* (1657).

William Smith (1769–1839) drew some of the first geological maps and began the process of ordering rock strata (layers) by examining the fossils contained in them.

James Hutton is often viewed as the first modern geologist. In 1785 he presented a paper entitled *Theory of the Earth* to the Royal Society of Edinburgh. In his paper, he explained his theory that the Earth must be much older than had previously been supposed in order to allow enough time for mountains to be eroded and for sediments to form new rocks at the bottom of the sea, which in turn were raised up to become dry land. Hutton published a two-volume version of his ideas in 1795 (Vol. 1, Vol. 2).

Scotsman James Hutton, father of modern geology

Followers of Hutton were known as *Plutonists* because they believed that some rocks were formed by vulcanism, which is the deposition of lava from volcanoes, as opposed to the *Neptunists*, led by

Abraham Werner, who believed that all rocks had settled out of a large ocean whose level gradually dropped over time.

The first geological map of the U.S. was produced in 1809 by William Maclure. In 1807, Maclure commenced the self-imposed task of making a geological survey of the United States. Almost every state in the Union was traversed and mapped by him, the Allegheny Mountains being crossed and recrossed some 50 times. The results of his unaided labours were submitted to the American Philosophical Society in a memoir entitled *Observations on the Geology of the United States explanatory of a Geological Map,* and published in the Society's Transactions, together with the nation's first geological map. This antedates William Smith's geological map of England by six years, although it was constructed using a different classification of rocks.

Sir Charles Lyell first published his famous book, *Principles of Geology,* in 1830. This book, which influenced the thought of Charles Darwin, successfully promoted the doctrine of uniformitarianism. This theory states that slow geological processes have occurred throughout the Earth's history and are still occurring today. In contrast, catastrophism is the theory that Earth's features formed in single, catastrophic events and remained unchanged thereafter. Though Hutton believed in uniformitarianism, the idea was not widely accepted at the time.

Much of 19th-century geology revolved around the question of the Earth's exact age. Estimates varied from a few hundred thousand to billions of years. By the early 20th century, radiometric dating allowed the Earth's age to be estimated at two billion years. The awareness of this vast amount of time opened the door to new theories about the processes that shaped the planet.

Some of the most significant advances in 20th-century geology have been the development of the theory of plate tectonics in the 1960s and the refinement of estimates of the planet's age. Plate tectonics theory arose from two separate geological observations: seafloor spreading and continental drift. The theory revolutionized the Earth sciences. Today the Earth is known to be approximately 4.5 billion years old.

Fields or Related Disciplines

- Earth science

- Earth system science

- Economic geology

    o Mining geology

    o Petroleum geology

- Engineering geology

- Environmental geology

- Environmental science

- Geoarchaeology

- Geochemistry
  - o Biogeochemistry
  - o Isotope geochemistry
- Geochronology
- Geodetics
- Geography
- Geological modelling
- Geometallurgy
- Geomicrobiology
- Geomorphology
- Geomythology
- Geophysics
- Glaciology
- Historical geology
- Hydrogeology
- Meteorology
- Mineralogy
- Oceanography
  - o Marine geology
- Paleoclimatology
- Paleontology
  - o Micropaleontology
  - o Palynology
- Petrology
- Petrophysics
- Physical geography
- Plate tectonics
- Sedimentology

- Seismology

- Soil science

    o Pedology (soil study)

- Speleology

- Stratigraphy

    o Biostratigraphy

    o Chronostratigraphy

    o Lithostratigraphy

- Structural geology

- Systems geology

- Volcanology

Planetary Geology

Planetary geologist and NASA astronaut Harrison "Jack" Schmitt collecting lunar samples during the Apollo 17 mission in early-December 1972

Planetary geology, alternatively known as astrogeology or exogeology, is a planetary science discipline concerned with the geology of the celestial bodies such as the planets and their moons, asteroids, comets, and meteorites. Although the geo- prefix typically indicates topics of or relating to the Earth, planetary geology is named as such for historical and convenience reasons; applying geological science to other planetary bodies. Due to the types of investigations involved, it is also closely linked with Earth-based geology.

Planetary geology includes such topics as determining the internal structure of the terrestrial planets, and also looks at planetary volcanism and surface processes such as impact craters, fluvial and aeolian processes. The structures of the giant planets and their moons are also examined, as is the make-up of the minor bodies of the Solar System, such as asteroids, the Kuiper Belt, and comets.

History of Planetary Geology

Eugene Shoemaker is credited with bringing geologic principles to planetary mapping and creating the branch of planetary science in the early 1960s, the Astrogeology Research Program, within the United States Geological Survey. He made important contributions to the field and the study of impact craters, Selenography (study of the Moon), asteroids, and comets.

Today many institutions are concerned with the study and communication of planetary sciences and planetary geology. The Visitor Center at Barringer Meteor Crater near Winslow, Arizona includes a Museum of planetary geology. The Geological Society of America's Planetary Geology Division has been growing and thriving since May 1981 and has two mottos: "One planet just isn't enough!" and ""The GSA Division with the biggest field area!"

Major centers for planetary science research include the Lunar and Planetary Institute, the Applied Physics Laboratory, the Planetary Science Institute, the Jet Propulsion Laboratory, Southwest Research Institute, and Johnson Space Center. Additionally, several universities conduct extensive planetary science research, including Montana State University, Brown University, the University of Arizona, Caltech, the University of Colorado, Western Michigan University, MIT, and Washington University in St. Louis.

Features and Terms

Planetary geology uses a wide variety of standardised descriptor names for features. All planetary feature names recognised by the International Astronomical Union combine one of these names with a possibly unique identifying name. The conventions which decide the more precise name are dependent on which planetary body the feature is on, but the standard descriptors are in general common to all astronomical planetary bodies. Some names have a long history of historical usage, but new must be recognised by the IAU Working Group for Planetary System Nomenclature as features are mapped and described by new planetary missions. This means that in some cases names may change as new imagery becomes available, or in other cases widely adopted informal names changed in line with the rules. The standard names are chosen to consciously avoid interpreting the underlying cause of the feature, but rather to describe only its appearance.

Feature	Pronunciation	Description	Designation
Albedo feature	/ælˈbiːdoʊ/	An area which shows a contrast in brightness or darkness (albedo) with adjacent areas. This term is implicit.	AL
Arcus, arcūs	/ˈɑːrkəs/	Arc: curved feature	AR
Astrum, astra	/ˈæstrəm/, /ˈæstrə/	Radial-patterned features on Venus	AS
Catena, catenae	/kəˈtiːnə/, /kəˈtiːniː/	A chain of craters e.g. Enki Catena.	CA

Cavus, cavi	/ˈkeɪvəs/, /ˈkeɪvaɪ/	Hollows, irregular steep-sided depressions usually in arrays or clusters	CB
Chaos	/ˈkeɪ.ɒs/	A distinctive area of broken or jumbled terrain e.g. Iani Chaos.	CH
Chasma, chasmata	/ˈkæzmə/, /ˈkæzmətə/	Deep, elongated, steep-sided depression e.g. Eos Chasma.	CM
Colles	/ˈkɒliːz/	A collection of small hills or knobs.	CO
Corona, coronae	/kɒˈroʊnə/, /kɒˈroʊniː/	An oval feature. Used only on Venus and Miranda.	CR
Crater, craters	/ˈkreɪtər/	A circular depression likely created by impact event. This term is implicit.	AA
Dorsum, dorsa	/ˈdɔːrsəm/, /ˈdɔːrsə/	Ridge, sometimes called a wrinkle ridge e.g. Dorsum Buckland.	DO
Eruptive center		An active volcano on Io. This term is implicit.	ER
Facula, faculae	/ˈfækjʊlə/, /ˈfækjʊliː/	Bright spot	FA
Farrum, farra	/ˈfærəm/, /ˈfærə/	Pancake-like structure, or a row of such structures. Used only on Venus.	FR
Flexus, flexūs	/ˈflɛksəs/	Very low curvilinear ridge with a scalloped pattern	FE
Fluctus, fluctūs	/ˈflʌktəs/	Terrain covered by outflow of liquid. Used on Venus, Io and Titan.	FL
Flumen, flumina	/ˈfluːmɪn/, /ˈfluːmɪnə/	Channel on Titan that might carry liquid	FM
Fossa, fossae	/ˈfɒsə/, /ˈfɒsiː/	Long, narrow, shallow depression	FO
Fretum, freta	/ˈfriːtəm/, /ˈfriːtə/	Strait of liquid connecting two larger areas of liquid. Used only on Titan.	FT
Insula, insulae	/ˈɪnsjuːlə/, /ˈɪnsjuːliː/	Island (islands), an isolated land area (or group of such areas) surrounded by, or nearly surrounded by, a liquid area (sea or lake). Used only on Titan.	IN
Labes, labes	/ˈleɪbiːz/	Landslide debris. Used only on Mars.	LA
Labyrinthus, labyrinthi	/læbɪˈrɪnθəs/, /læbɪˈrɪnθaɪ/	Complex of intersecting valleys or ridges.	LB
Lacuna, lacunae	/ləˈkjuːnə/, /ləˈkjuːniː/	Irregularly shaped depression having the appearance of a dry lake bed. Used only on Titan.	LU
Lacus, lacūs	/ˈleɪkəs/	A "lake" or small plain on Moon and Mars; on Titan, a "lake" or small, dark plain with discrete, sharp boundaries.	LC
Landing site name		Lunar features at or near Apollo landing sites	LF
Large ringed feature		Cryptic ringed features	LG
Lenticula, lenticulae	/lɛnˈtɪkjʊlə/, /lɛnˈtɪkjʊliː/	Small dark spots on Europa	LE
Linea, lineae	/ˈlɪniːə/, /ˈlɪniː.iː/	Dark or bright elongate marking, may be curved or straight	LI
Macula, maculae	/ˈmækjʊlə/, /ˈmækjʊliː/	Dark spot, may be irregular	MA
Mare, maria	/ˈmɑːriː/ ~ /ˈmɑːreɪ/, /ˈmɑːriə/	A "sea" or large circular plain on Moon and Mars, e.g. Mare Erythraeum; on Titan, large expanses of dark materials thought to be liquid hydrocarbons, e.g. Ligeia Mare.	ME
Mensa, mensae	/ˈmɛnsə/, /ˈmɛnsiː/	A flat-topped prominence with cliff-like edges, i.e. a mesa.	MN

Mons, montes	/ˈmɒnz/, /ˈmɒntiːz/	Mons refers to a mountain. Montes refers to a mountain range.	MO
Oceanus	/oʊʃiˈaɪnəs/	Very large dark area. The only feature with this designation is Oceanus Procellarum.	OC
Palus, paludes	/ˈpeɪləs/, /pəˈljuːdiːz/	"Swamp"; small plain. Used on the Moon and Mars.	PA
Patera, paterae	/ˈpætərə/, /ˈpætəriː/	Irregular crater, or a complex one with scalloped edges e.g. Ah Peku Patera. Usually refers to the dish-shaped depression atop a volcano.	PE
Planitia, planitiae	/pləˈnɪʃə/, /pləˈnɪʃiː/	Low plain e.g. Amazonis Planitia.	PL
Planum, plana	/ˈpleɪnəm/, /ˈpleɪnə/	A plateau or high plain e.g. Planum Boreum.	PM
Plume		A cryovolcanic feature on Triton. This term is currently unused.	PU
Promontorium, promontoria	/prɒmənˈtɔəriəm/, /prɒmənˈtɔəriə/	"Cape"; headland. Used only on the Moon.	PR
Regio, regiones	/ˈriːdʒioʊ/ ~ /ˈrɛdʒioʊ/, /rɛdʒiˈoʊniːz/	Large area marked by reflectivity or color distinctions from adjacent areas, or a broad geographic region	RE
Reticulum, reticula	/rɪˈtɪkjʊləm/, /rɪˈtɪkjʊlə/	reticular (netlike) pattern on Venus	RT
Rima, rimae	/ˈraɪmə/, /ˈraɪmiː/	Fissure. Used only on the Moon.	RI
Rupes, rupes	/ˈruːpiːz/	Scarp	RU
Satellite feature		A feature that shares the name of an associated feature, for example Hertzsprung D.	SF
Scopulus, scopuli	/ˈskɒpjʊlə/, /ˈskɒpjʊlaɪ/	Lobate or irregular scarp	SC
Serpens, serpentes	/ˈsɜːrpɛnz/, /sərˈpɛntiːz/	Sinuous feature with segments of positive and negative relief along its length	SE
Sinus	/ˈsaɪnəs/	"Bay"; small plain on Moon or Mars, e.g. Sinus Meridiani; On Titan, bay within bodies of liquid.	SI
Sulcus, sulci	/ˈsʌlkəs/, /ˈsʌlsaɪ/	Subparallel furrows and ridges	SU
Terra, terrae	/ˈtɛrə/, /ˈtɛriː/	Extensive land mass e.g. Arabia Terra, Aphrodite Terra.	TA
Tessera, tesserae	/ˈtɛsərə/, /ˈtɛsəriː/	An area of tile-like, polygonal terrain. This term is used only on Venus.	TE
Tholus, tholi	/ˈθoʊləs/, /ˈθoʊlaɪ/	Small domical mountain or hill e.g. Hecates Tholus.	TH
Undae	/ˈʌndiː/	A field of dunes. Used on Venus, Mars and Titan.	UN
Vallis, valles	/ˈvælɪs/, /ˈvæliːz/	A valley e.g. Valles Marineris.	VA
Vastitas, vastitates	/ˈvæstɪtəs/, /væstɪˈteɪtiːz/	An extensive plain. The only feature with this designation is Vastitas Borealis.	VS
Virga, virgae	/ˈvɜːrgə/, /ˈvɜːrdʒiː/	A streak or stripe of color. This term is currently used only on Titan.	VI

By Planet

- Geological features of the solar system
- Geological history of Earth
- Geology of Mercury
- Geology of Venus
- Geology of the Moon
- Geology of Mars
- Geology of Vesta
- Geology of Ceres
- Geology of Callisto
- Geology of Europa
- Geology of Ganymede
- Geology of Io
- Geology of Titan
- Geology of Triton
- Geology of Pluto
- Geology of Charon

Climatology

Climatology *or* climate science is the study of climate, scientifically defined as weather conditions averaged over a period of time. This modern field of study is regarded as a branch of the atmospheric sciences and a subfield of phys-ical geography, which is one of the Earth sciences. Climatology now includes aspects of oceanog-raphy and biogeochemistry. Basic knowledge of climate can be used within shorter term weather forecasting using analog techniques such as the El Niño–Southern Oscillation (ENSO), the Madden–Julian oscillation (MJO), the North Atlantic oscillation (NAO), the Northern Annular Mode (NAM) which is also known as the Arctic oscillation (AO), the Northern Pacific (NP) Index, the Pacific decadal oscillation (PDO), and the Interdecadal Pacific Oscillation (IPO). Climate models are used for a variety of purposes from study of the dynamics of the weather and climate system to projections of future climate.

History

Chinese scientist Shen Kuo (1031–1095) inferred that climates naturally shifted over an enormous span of time, after observing petrified bamboos found underground near Yanzhou (modern day Yan'an, Shaanxi province), a dry-climate area unsuitable for the growth of bamboo.

Early climate researchers include Edmund Halley, who published a map of the trade winds in 1686 after a voyage to the southern hemisphere. Benjamin Franklin (1706–1790) first mapped the course of the Gulf Stream for use in sending mail from the United States to Europe. Francis Galton (1822–1911) invented the term *anticyclone*. Helmut Landsberg (1906–1985) fostered the use of statistical analysis in climatology, which led to its evolution into a physical science.

Different Approaches

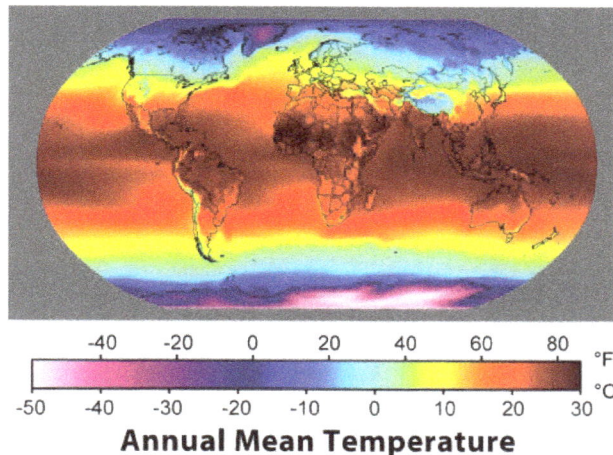

Annual Mean Temperature

Map of the average temperature over 30 years. Data sets formed from the long-term average of historical weather parameters are sometimes called a "climatology".

Climatology is approached in a variety of ways. Paleoclimatology seeks to reconstruct past climates by examining records such as ice cores and tree rings (dendroclimatology). Paleotempestology uses these same records to help determine hurricane frequency over millennia. The study of contemporary climates incorporates meteorological data accumulated over many years, such as records of rainfall, temperature and atmospheric composition. Knowledge of the atmosphere and its dynamics is also embodied in models, either statistical or mathematical, which help by integrating different observations and testing how they fit together. Modeling is used for understanding past, present and potential future climates. Historical climatology is the study of climate as related to human history and thus focuses only on the last few thousand years.

Climate research is made difficult by the large scale, long time periods, and complex processes which govern climate. Climate is governed by physical laws which can be expressed as differential equations. These equations are coupled and nonlinear, so that approximate solutions are obtained by using numerical methods to create global climate models. Climate is sometimes modeled as a stochastic process but this is generally accepted as an approximation to processes that are otherwise too complicated to analyze.

Indices

Scientists use climate indices based on several climate patterns (known as modes of variability) in their attempt to characterize and understand the various climate mechanisms that culminate in our daily weather. Much in the way the Dow Jones Industrial Average, which is based on the stock prices of 30 companies, is used to represent the fluctuations in the stock market as a whole, climate indices are used to represent the essential elements of climate. Climate indices are generally devised with the twin objectives of simplicity and completeness, and each index typically represents the status and timing of the climate factor it represents. By their very nature, indices are simple, and combine many details into a generalized, overall description of the atmosphere or ocean which can be used to characterize the factors which impact the global climate system.

El Niño–Southern Oscillation

WARM EPISODE RELATIONSHIPS   DECEMBER - FEBRUARY

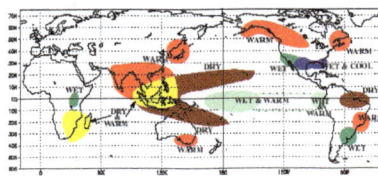

WARM EPISODE RELATIONSHIPS   JUNE - AUGUST

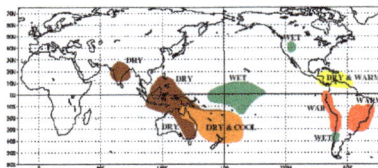

El Niño impacts

COLD EPISODE RELATIONSHIPS   DECEMBER - FEBRUARY

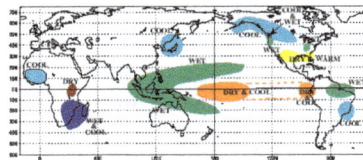

COLD EPISODE RELATIONSHIPS   JUNE - AUGUST

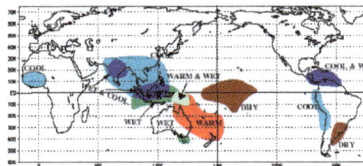

La Niña impacts

El Niño–Southern Oscillation (ENSO) is a global coupled ocean-atmosphere phenomenon. The Pacific ocean signatures, El Niño and La Niña are important temperature fluctuations in surface waters of the tropical Eastern Pacific Ocean. The name El Niño, from the Spanish for "the little boy", refers to the Christ child, because the phenomenon is usually noticed around Christmas time in the Pacific Ocean off the west coast of South America. La Niña means "the little girl". Their effect on climate in the subtropics and the tropics are profound. The atmospheric signature, the Southern Oscillation (SO) reflects the monthly or seasonal fluctuations in the air pressure difference between Tahiti and Darwin. The most recent occurrence of El Niño started in September 2006 and lasted until early 2007.

ENSO is a set of interacting parts of a single global system of coupled ocean-atmosphere climate fluctuations that come about as a consequence of oceanic and atmospheric circulation. ENSO is the most prominent known source of inter-annual variability in weather and climate around the world. The cycle occurs every two to seven years, with El Niño lasting nine months to two years within the longer term cycle, though not all areas globally are affected. ENSO has signatures in the Pacific, Atlantic and Indian Oceans.

In the Pacific, during major warm events, El Niño warming extends over much of the tropical Pacific and becomes clearly linked to the SO intensity. While ENSO events are basically in phase between the Pacific and Indian Oceans, ENSO events in the Atlantic Ocean lag behind those in the Pacific by 12–18 months. Many of the countries most affected by ENSO events are developing countries within tropical sections of continents with economies that are largely dependent upon their agricultural and fishery sectors as a major source of food supply, employment, and foreign exchange. New capabilities to predict the onset of ENSO events in the three oceans can have global socio-economic impacts. While ENSO is a global and natural part of the Earth's climate, whether its intensity or frequency may change as a result of global warming is an important concern. Low-frequency variability has been evidenced: the quasi-decadal oscillation (QDO). Inter-decadal (ID) modulation of ENSO (from PDO or IPO) might exist. This could explain the so-called protracted ENSO of the early 1990s.

Madden–Julian Oscillation

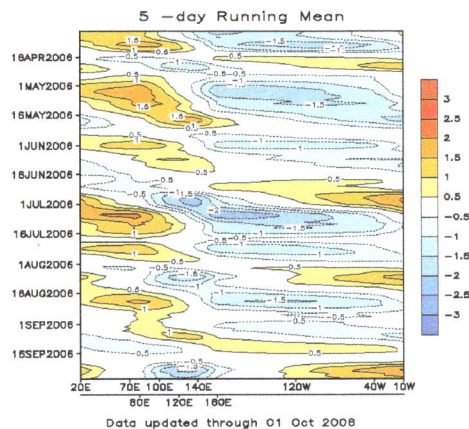

Note how the MJO moves eastward with time.

The Madden–Julian oscillation (MJO) is an equatorial traveling pattern of anomalous rainfall that is planetary in scale. It is characterized by an eastward progression of large regions of both enhanced and suppressed tropical rainfall, observed mainly over the Indian and Pacific Oceans. The anomalous rainfall is usually first evident over the western Indian Ocean, and remains evident as it propagates over the very warm ocean waters of the western and central tropical Pacific. This pattern of tropical rainfall then generally becomes very nondescript as it moves over the cooler ocean waters of the eastern Pacific but reappears over the tropical Atlantic and Indian Oceans. The wet phase of enhanced convection and precipitation is followed by a dry phase where convection is suppressed. Each cycle lasts approximately 30–60 days. The MJO is also known as the 30- to 60-day oscillation, 30- to 60-day wave, or the intraseasonal oscillation.

North Atlantic Oscillation (NAO)

Indices of the NAO are based on the difference of normalized sea level pressure (SLP) between Ponta Delgada, Azores and Stykkisholmur/Reykjavik, Iceland. The SLP anomalies at each station were normalized by division of each seasonal mean pressure by the long-term mean (1865–1984) standard deviation. Normalization is done to avoid the series of being dominated by the greater variability of the northern of the two stations. Positive values of the index indicate stronger-than-average westerlies over the middle latitudes.

Northern Annular Mode (NAM) or Arctic Oscillation (AO)

The NAM, or AO, is defined as the first EOF of northern hemisphere winter SLP data from the tropics and subtropics. It explains 23% of the average winter (December–March) variance, and it is dominated by the NAO structure in the Atlantic. Although there are some subtle differences from the regional pattern over the Atlantic and Arctic, the main difference is larger amplitude anomalies over the North Pacific of the same sign as those over the Atlantic. This feature gives the NAM a more annular (or zonally symmetric) structure.

Northern Pacific (NP) Index

The NP Index is the area-weighted sea level pressure over the region 30N–65N, 160E–140W.

Pacific Decadal Oscillation (PDO)

The PDO is a pattern of Pacific climate variability that shifts phases on at least inter-decadal time scale, usually about 20 to 30 years. The PDO is detected as warm or cool surface waters in the Pacific Ocean, north of 20° N. During a "warm", or "positive", phase, the west Pacific becomes cool and part of the eastern ocean warms; during a "cool" or "negative" phase, the opposite pattern occurs. The mechanism by which the pattern lasts over several years has not been identified; one suggestion is that a thin layer of warm water during summer may shield deeper cold waters. A PDO signal has been reconstructed to 1661 through tree-ring chronologies in the Baja California area.

Interdecadal Pacific Oscillation (IPO)

The Interdecadal Pacific oscillation (IPO or ID) display similar sea surface temperature (SST) and sea level pressure patterns to the PDO, with a cycle of 15–30 years, but affects both the north and south Pacific. In the tropical Pacific, maximum SST anomalies are found away from the equator. This is quite different from the quasi-decadal oscillation (QDO) with a period of 8–12 years and maximum SST anomalies straddling the equator, thus resembling ENSO.

Models

Climate models use quantitative methods to simulate the interactions of the atmosphere, oceans, land surface, and ice. They are used for a variety of purposes from study of the dynamics of the weather and climate system to projections of future climate. All climate models balance, or very nearly balance, incoming energy as short wave (including visible) electromagnetic radiation to the earth with outgoing energy as long wave (infrared) electromagnetic radiation from the earth. Any unbalance results in a change in the average temperature of the earth.

The most talked-about models of recent years have been those relating temperature to emissions of carbon dioxide. These models predict an upward trend in the surface tem-perature record, as well as a more rapid increase in temperature at higher latitudes.

Models can range from relatively simple to quite complex:

- A simple radiant heat transfer model that treats the earth as a single point and averages outgoing energy

- this can be expanded vertically (radiative-convective models), or horizontally

- finally, (coupled) atmosphere–ocean–sea ice global climate models discretise and solve the full equations for mass and energy transfer and radiant exchange.

Differences with Meteorology

In contrast to meteorology, which focuses on short term weather systems lasting up to a few weeks, climatology studies the frequency and trends of those systems. It studies the periodicity of weather events over years to millennia, as well as changes in long-term average weather patterns, in rela-tion to atmospheric conditions. Climatologists study both the nature of climates – local, regional or global – and the natural or human-induced factors that cause climates to change. Climatology considers the past and can help predict future climate change.

Phenomena of climatological interest include the atmospheric boundary layer, circulation pat-terns, heat transfer (radiative, convective and latent), interactions between the atmosphere and the oceans and land surface (particularly vegetation, land use and topography), and the chemical and physical composition of the atmosphere.

Use in Weather Forecasting

A more complicated way of making a forecast, the analog technique requires remembering a pre-vious weather event which is expected to be mimicked by an upcoming event. What makes it a difficult technique to use is that there is rarely a perfect analog for an event in the future. Some call this type of forecasting pattern recognition, which remains a useful method of observing rainfall over data voids such as oceans with knowledge of how satellite imagery relates to precipitation rates over land, as well as the forecasting of precipitation amounts and distribution in the future. A variation on this theme is used in Medium Range forecasting, which is known as teleconnections, when you use systems in other locations to help pin down the location of another system within the surrounding regime. One method of using teleconnections are by using climate indices such as ENSO-related phenomena.

Meteorology

Meteorology is the interdisciplinary scientific study of the atmosphere. The study of meteorology dates back millennia, though significant progress in meteorology did not occur until the 18th century. The 19th century saw modest progress in the field after weather observation networks were formed across broad regions. Prior attempts at prediction of weather depended on historical data. It wasn't until after the elucidation of the laws of physics and, more particularly, the development of the computer, allowing for the automated solution of the great many equations that model the weather, in the latter half of the 20th century that significant breakthroughs in weather forecasting were achieved.

Meteorological phenomena are observable weather events that are explained by the science of meteorology. Meteorological phenomena are described and quantified by the variables of Earth's atmosphere: temperature, air pressure, water vapor, mass flow, and the variations and interactions of those variables, and how they change over time. Different spatial scales are used to describe and predict weather on local, regional, and global levels.

Meteorology, climatology, atmospheric physics, and atmospheric chemistry are sub-disciplines of the atmospheric sciences. Meteorology and hydrology compose the interdisciplinary field of hydrometeorology. The interactions between Earth's atmosphere and its oceans are part of a coupled ocean-atmosphere system. Meteorology has application in many diverse fields such as the military, energy production, transport, agriculture, and construction.

History

Parhelion (sundog) at Savoie

The beginnings of meteorology can be traced back to ancient India, as the Upanishads contain serious discussion about the processes of cloud formation and rain and the seasonal cycles caused by the movement of Earth around the sun. Varāhamihira's classical work *Brihatsamhita,* written about 500 AD, provides clear evidence that a deep knowledge of atmospheric processes existed even in those times.

In 350 BC, Aristotle wrote *Meteorology*. Aristotle is considered the founder of meteorology. One of the most impressive achievements described in the *Meteorology* is the description of what is now known as the hydrologic cycle.

The book De Mundo (composed before 250 BC or between 350 and 200 BC) noted

> If the flashing body is set on fire and rushes violently to the Earth it is called a thunderbolt; if it be only half of fire, but violent also and massive, it is called a *meteor;* if it is entirely free from fire, it is called a smoking bolt. They are all called 'swooping bolts', because they swoop down upon the Earth. Lightning is sometimes smoky, and is then called 'smouldering lightning"; sometimes it darts quickly along, and is then said to be *vivid. At other times, it travels in crooked lines, and is called forked lightning.* When it swoops down upon some object it is called 'swooping lightning'.

The Greek scientist Theophrastus compiled a book on weather forecasting, called the *Book of Signs*. The work of Theophrastus remained a dominant influence in the study of weather and in weather forecasting for nearly 2,000 years. In 25 AD, Pomponius Mela, a geographer for the Roman Empire, formalized the climatic zone system. According to Toufic Fahd, around the 9th century, Al-Dinawari wrote the Kitab al-Nabat (Book of Plants), in which he deals with the application of meteorology to agriculture during the Muslim Agricultural Revolution. He describes the meteorological character of the sky, the planets and constellations, the sun and moon, the lunar phases indicating seasons and rain, the anwa (heavenly bodies of rain), and atmospheric phenomena such as winds, thunder, lightning, snow, floods, valleys, rivers, lakes.

Research of Visual Atmospheric Phenomena

Twilight at Baker Beach

Ptolemy wrote on the atmospheric refraction of light in the context of astronomical observations. In 1021, Alhazen showed that atmospheric refraction is also responsible for twilight; he estimated that twilight begins when the sun is 19 degrees below the horizon, and also used a geometric determination based on this to estimate the maximum possible height of the Earth's atmosphere as 52,000 *passuum* (about 49 miles, or 79 km).

St. Albert the Great was the first to propose that each drop of falling rain had the form of a small sphere, and that this form meant that the rainbow was produced by light interacting with each raindrop. Roger Bacon was the first to calculate the angular size of the rainbow. He stated that a rainbow summit can not appear higher than 42 degrees above the horizon. In the late 13th century and early 14th century, Kamāl al-Dīn al-Fārisī and Theodoric of Freiberg were the first to give the correct explanations for the primary rainbow phenomenon. Theoderic went further and also explained the secondary rainbow. In 1716, Edmund Halley suggested that aurorae are caused by "magnetic effluvia" moving along the Earth's magnetic field lines.

Instruments and Classification Scales

THE ROBINSON ANEMOMETER.

A hemispherical cup anemometer

In 1441, King Sejong's son, Prince Munjong, invented the first standardized rain gauge. These were sent throughout the Joseon Dynasty of Korea as an official tool to assess land taxes based upon a farmer's potential harvest. In 1450, Leone Battista Alberti developed a swinging-plate anemometer, and was known as the first *anemometer*. In 1607, Galileo Galilei constructed a thermoscope. In 1611, Johannes Kepler wrote the first scientific treatise on snow crystals: "Strena Seu de Nive Sexangula (A New Year's Gift of Hexagonal Snow)". In 1643, Evangelista Torricelli invented the mercury barometer. In 1662, Sir Christopher Wren invented the mechanical, self-emptying, tipping bucket rain gauge. In 1714, Gabriel Fahrenheit created a reliable scale for measuring temperature with a mercury-type thermometer. In 1742, Anders Celsius, a Swedish astronomer, proposed the "centigrade" temperature scale, the predecessor of the current Celsius scale. In 1783, the first hair hygrometer was demonstrated by Horace-Bénédict de Saussure. In 1802–1803, Luke Howard wrote *On the Modification of Clouds, in which he assigns* cloud types Latin names. In 1806, Francis Beaufort introduced his system for classifying wind speeds. Near the end of the 19th century the first cloud atlases were published, including the *International Cloud Atlas,* which has remained in print ever since. The April 1960 launch of the first successful weather satellite, TIROS-1, marked the beginning of the age where weather information became available globally.

Atmospheric Composition Research

In 1648, Blaise Pascal rediscovered that atmospheric pressure decreases with height, and deduced that there is a vacuum above the atmosphere. In 1738, Daniel Bernoulli published *Hydrodynamics, initiating the* Kinetic theory of gases and established the basic laws for the theory of gases. In 1761, Joseph Black discovered that ice absorbs heat without changing its temperature when melting. In 1772, Black's student Daniel Rutherford discovered nitrogen, which he called *phlogisticated air, and together they developed the* phlogiston theory. In 1777, Antoine Lavoisier discovered oxygen and developed an explanation for combustion. In 1783, in Lavoisier's essay "Reflexions sur le phlogistique", he deprecates the phlogiston theory and proposes a caloric theory. In 1804, Sir John Leslie observed that a matte black surface radiates heat more effectively than a polished surface, suggesting the importance of black body radiation. In 1808, John Dalton defended caloric theory in *A New System of Chemistry and described how it combines with matter, especially gases; he proposed that the* heat capacity of gases varies inversely with atomic weight. In 1824, Sadi Carnot analyzed the efficiency of steam engines using caloric theory; he developed the notion of a reversible process and, in postulating that no such thing exists in nature, laid the foundation for the second law of thermodynamics.

Research Into Cyclones and Air Flow

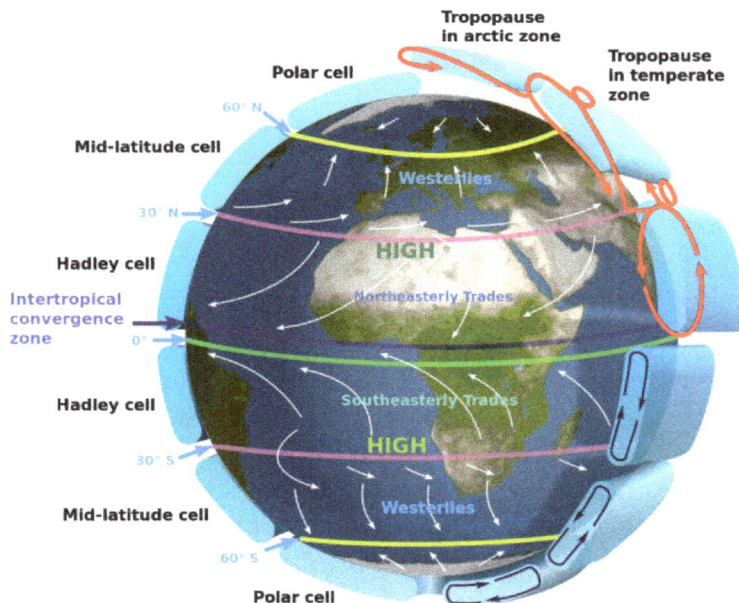

General Circulation of the Earth's Atmosphere: The westerlies and trade winds are part of the Earth's atmospheric circulation

In 1494, Christopher Columbus experienced a tropical cyclone, which led to the first written European account of a hurricane. In 1686, Edmund Halley presented a systematic study of the trade winds and monsoons and identified solar heating as the cause of atmospheric motions. In 1735, an *ideal explanation of* global circulation through study of the trade winds was written by George Hadley. In 1743, when Benjamin Franklin was prevented from seeing a lunar eclipse by a hurricane, he decided that cyclones move in a contrary manner to the winds at their periphery. Understanding the kinematics of how exactly the rotation of the Earth affects airflow was partial at first. Gaspard-Gustave Coriolis published a paper in 1835 on the energy yield of machines with rotating parts, such as waterwheels. In 1856, William Ferrel proposed the existence of a circulation cell in the mid-latitudes, and the air within deflected by the Coriolis force resulting in the prevailing westerly winds. Late in the 19th century, the motion of air masses along isobars was understood to be the result of the large-scale interaction of the pressure gradient force and the deflecting force. By 1912, this deflecting force was named the Coriolis effect. Just after World War I, a group of meteorologists in Norway led by Vilhelm Bjerknes developed the Norwegian cyclone model that explains the generation, intensification and ultimate decay (the life cycle) of mid-latitude cyclones, and introduced the idea of fronts, that is, sharply defined boundaries between air masses. The group included Carl-Gustaf Rossby (who was the first to explain the large scale atmospheric flow in terms of fluid dynamics), Tor Bergeron (who first determined how rain forms) and Jacob Bjerknes.

Observation Networks and Weather Forecasting

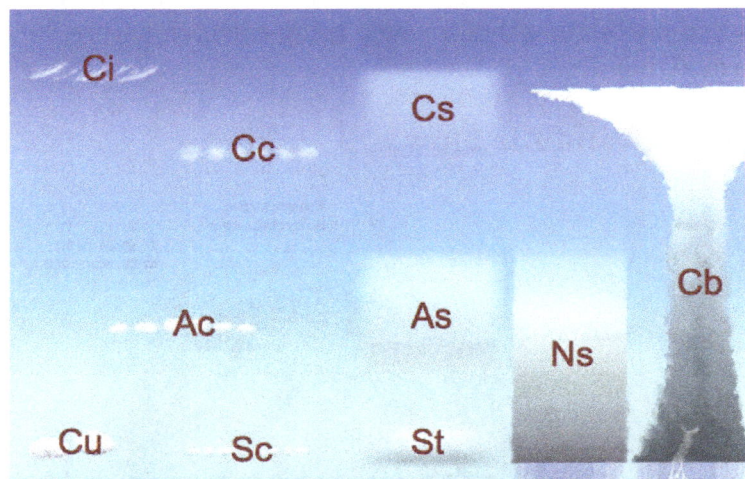

Cloud classification by altitude of occurrence

In 1654, Ferdinando II de Medici established the first *weather observing network, that consisted of meteorological stations in* Florence, Cutigliano, Vallombrosa, Bologna, Parma, Milan, Innsbruck, Osnabrück, Paris and Warsaw. The collected data were sent to Florence at regular time intervals. In 1832, an electromagnetic telegraph was created by Baron Schilling. The arrival of the electrical telegraph in 1837 afforded, for the first time, a practical method for quickly gathering surface weather observations from a wide area. This data could be used to produce maps of the state of the atmosphere for a region near the Earth's surface and to study how these states evolved through time. To make frequent weather forecasts based on these data required a reliable network of observations, but it was not until 1849 that the Smithsonian Institution began to establish an observation network across the United States under the leadership of Joseph Henry. Similar observation networks were established in Europe at this time. The Reverend William Clement Ley was key in understanding of cirrus clouds and early understandings of Jet Streams. Later after this Charles Kenneth Mackinnon Douglas known as 'CKM' Douglas read Ley's papers after his death and carried on the early study of weather systems. Nineteenth century researchers in meteorology were drawn from military or medical backgrounds, rather than trained as dedicated scientists. In 1854, the United Kingdom government appointed Robert FitzRoy to the new office of *Meteorological Statist to the Board of Trade* with the task of gathering weather observations at sea. FitzRoy's office became the United Kingdom Meteorological Office in 1854, the first national meteorological service in the world. The first daily weather forecasts made by FitzRoy's Office were published in *The Times* newspaper in 1860. The following year a system was introduced of hoisting storm warning cones at principal ports when a gale was expected.

Over the next 50 years many countries established national meteorological services. The India Meteorological Department (1875) was established to follow tropical cyclone and monsoon. The Finnish Meteorological Central Office (1881) was formed from part of Magnetic Observatory of Helsinki University. Japan's Tokyo Meteorological Observatory, the forerunner of the Japan Meteorological Agency, began constructing surface weather maps in 1883. The United States Weather Bureau (1890) was established under the United States Department of Agriculture. The Australian Bureau of Meteorology (1906) was established by a Meteorology Act to unify existing state meteorological services.

Numerical Weather Prediction

A meteorologist at the console of the IBM 7090 in the Joint Numerical Weather Prediction Unit. c. 1965

In 1904, Norwegian scientist Vilhelm Bjerknes first argued in his paper *Weather Forecasting as a Problem in Mechanics and Physics* that it should be possible to forecast weather from calculations based upon natural laws.

It was not until later in the 20th century that advances in the understanding of atmospheric physics led to the foundation of modern numerical weather prediction. In 1922, Lewis Fry Richardson published "Weather Prediction By Numerical Process", after finding notes and derivations he worked on as an ambulance driver in World War I. He described how small terms in the prognostic fluid dynamics equations that govern atmospheric flow could be neglected, and a numerical calculation scheme that could be devised to allow predictions. Richardson envisioned a large auditorium of thousands of people performing the calculations. However, the sheer number of calculations required was too large to complete without electronic computers, and the size of the grid and time steps used in the calculations led to unrealistic results. Though numerical analysis later found that this was due to numerical instability.

Starting in the 1950s, numerical forecasts with computers became feasible. The first weather forecasts derived this way used barotropic (single-vertical-level) models, and could successfully predict the large-scale movement of midlatitude Rossby waves, that is, the pattern of atmospheric lows and highs. In 1959, the UK Meteorological Office received its first computer, a Ferranti Mercury.

In the 1960s, the chaotic nature of the atmosphere was first observed and mathematically described by Edward Lorenz, founding the field of chaos theory. These advances have led to the current use of ensemble forecasting in most major forecasting centers, to take into account uncertainty arising from the chaotic nature of the atmosphere. Mathematical models used to predict the long term weather of the Earth (Climate models), have been developed that have a resolution today that are as coarse as the older weather prediction models. These climate models are used to investigate long-term climate shifts, such as what effects might be caused by human emission of greenhouse gases.

Meteorologists

Meteorologists are scientists who study meteorology. The American Meteorological Society published and continually updates an authoritative electronic *Meteorology Glossary*. Meteorologists work in government agencies, private consulting and research services, industrial enterprises, utilities, radio and television stations, and in education. In the United States, meteorologists held about 9,400 jobs in 2009.

Meteorologists are best known by the public for weather forecasting. Some radio and television weather forecasters are professional meteorologists, while others are reporters (weather specialist, weatherman, etc.) with no formal meteorological training. The American Meteorological Society and National Weather Association issue "Seals of Approval" to weather broadcasters who meet certain requirements.

Equipment

Each science has its own unique sets of laboratory equipment. In the atmosphere, there are many things or qualities of the atmosphere that can be measured. Rain, which can be observed, or seen anywhere and anytime was one of the first atmospheric qualities measured historically. Also, two other accurately measured qualities are wind and humidity. Neither of these can be seen but can be felt. The devices to measure these three sprang up in the mid-15th century and were respectively the rain gauge, the anemometer, and the hygrometer. Many attempts had been made prior to the 15th century to construct adequate equipment to measure the many atmospheric variables. Many were faulty in some way or were simply not reliable. Even Aristotle noted this in some of his work; as the difficulty to measure the air.

Sets of surface measurements are important data to meteorologists. They give a snapshot of a variety of weather conditions at one single location and are usually at a weather station, a ship or a weather buoy. The measurements taken at a weather station can include any number of atmospheric observables. Usually, temperature, pressure, wind measurements, and humidity are the variables that are measured by a thermometer, barometer, anemometer, and hygrometer, respectively. Professional stations may also include air quality sensors (carbon monoxide, carbon dioxide, methane, ozone, dust, and smoke), ceilometer (cloud ceiling), falling precipitation sensor, flood sensor, lightning sensor, microphone (explosions, sonic booms, thunder), pyranometer/pyrheliometer/spectroradiometer (IR/Vis/UV photodiodes), rain gauge/snow gauge, scintillation counter (background radiation, fallout, radon), seismometer (earthquakes and tremors), transmissometer (visibility), and a GPS clock for data logging. Upper air data are of crucial importance for weather forecasting. The most widely used technique is launches of radiosondes. Supplementing the radiosondes a network of aircraft collection is organized by the World Meteorological Organization.

Satellite image of Hurricane Hugo with a polar low visible at the top of the image.

Remote sensing, as used in meteorology, is the concept of collecting data from remote weather events and subsequently producing weather information. The common types of remote sensing are Radar, Lidar, and satellites (or photogrammetry). Each collects data about the atmosphere from a remote location and, usually, stores the data where the instrument is located. Radar and Lidar are not passive because both use EM radiation to illuminate a specific portion of the atmosphere. Weather satellites along with more general-purpose Earth-observing satellites circling the earth at various altitudes have become an indispensable tool for studying a wide range of phenomena from forest fires to El Niño.

Spatial Scales

In the study of the atmosphere, meteorology can be divided into distinct areas that depend on both time and spatial scales. At one extreme of this scale is climatology. In the timescales of hours to days, meteorology separates into micro-, meso-, and synoptic scale meteorology. Respectively, the geospatial size of each of these three scales relates directly with the appropriate timescale.

Other subclassifications are used to describe the unique, local, or broad effects within those sub-classes.

Typical Scales of Atmospheric Motion Systems	
Type of motion	Horizontal scale (meter)
Molecular mean free path	10^{-3}
Minute turbulent eddies	$10^{-2} - 10^{-1}$
Small eddies	$10^{-1} - 1$
Dust devils	$1 - 10$
Gusts	$10 - 10^2$
Tornadoes	10^2
Thunderclouds	10^3
Fronts, squall lines	$10^4 - 10^5$
Hurricanes	10^5
Synoptic Cyclones	10^6
Planetary waves	10^7
Atmospheric tides	10^7
Mean zonal wind	10^7

Microscale

Microscale meteorology is the study of atmospheric phenomena on a scale of about 1 kilometre (0.62 mi) or less. Individual thunderstorms, clouds, and local turbulence caused by buildings and other obstacles (such as individual hills) are modeled on this scale.

Mesoscale

Mesoscale meteorology is the study of atmospheric phenomena that has horizontal scales ranging from 1 km to 1000 km and a vertical scale that starts at the Earth's surface and includes the atmospheric boundary layer, troposphere, tropopause, and the lower section of the stratosphere. Mesoscale timescales last from less than a day to weeks. The events typically of interest are thunderstorms, squall lines, fronts, precipitation bands in tropical and extratropical cyclones, and topographically generated weather systems such as mountain waves and sea and land breezes.

Synoptic Scale

Synoptic scale meteorology predicts atmosperic changes at scales up to 1000 km and 105 sec (28 days), in time and space. At the synoptic scale, the Coriolis acceleration acting on moving air masses (outside of the tropics), plays a dominant role in predictions. The phenomena typically described by synoptic meteorology include events such as extratropical cyclones, baroclinic troughs and ridges, frontal zones, and to some extent jet streams. All of these are typically given on weather maps for a specific time. The minimum horizontal scale of synoptic phenomena is limited to the spacing between surface observation stations.

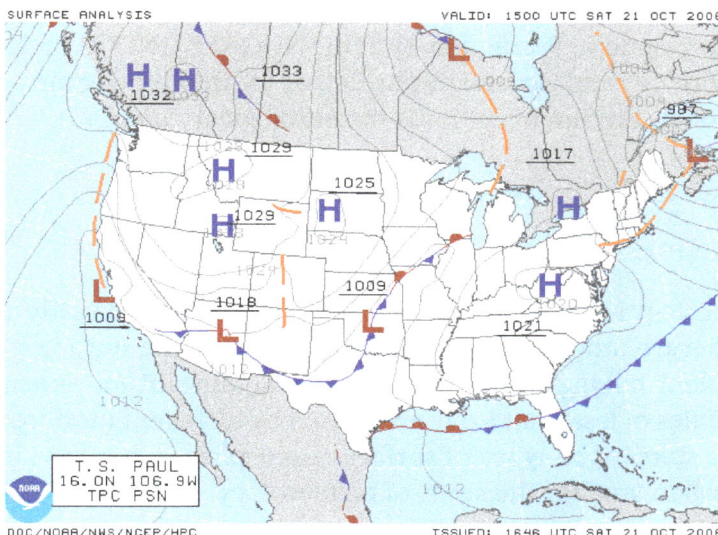

NOAA: Synoptic scale weather analysis.

Global Scale

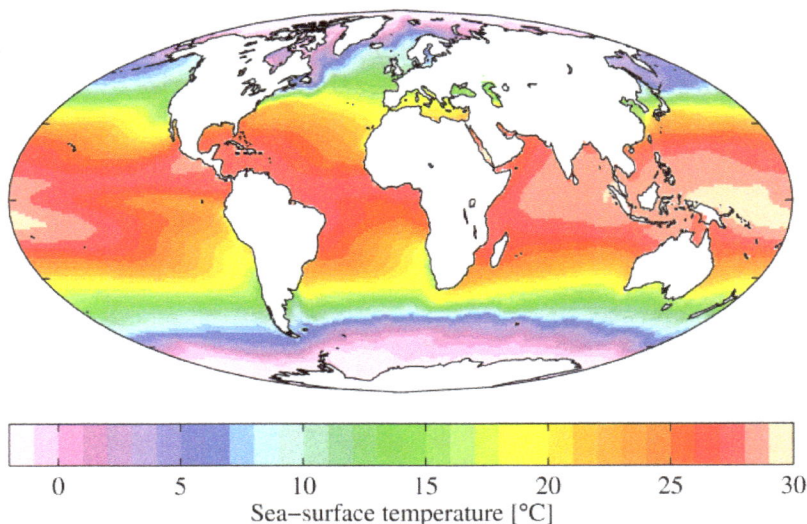

Annual mean sea surface temperatures.

Global scale meteorology is the study of weather patterns related to the transport of heat from the tropics to the poles. Very large scale oscillations are of importance at this scale. These oscillations have time periods typically on the order of months, such as the Madden–Julian oscillation, or years, such as the El Niño–Southern Oscillation and the Pacific decadal oscillation. Global scale meteorology pushes into the range of climatology. The traditional definition of climate is pushed into larger timescales and with the understanding of the longer time scale global oscillations, their effect on climate and weather disturbances can be included in the synoptic and mesoscale timescales predictions.

Numerical Weather Prediction is a main focus in understanding air–sea interaction, tropical meteorology, atmospheric predictability, and tropospheric/stratospheric processes. The Naval Research Laboratory in Monterey, California, developed a global atmospheric model called Navy Operational Global Atmospheric Prediction System (NOGAPS). NOGAPS is run operationally at Fleet Numerical Meteorology and Oceanography Center for the United States Military. Many other global atmospheric models are run by national meteorological agencies.

Some Meteorological Principles

Boundary Layer Meteorology

Boundary layer meteorology is the study of processes in the air layer directly above Earth's surface, known as the atmospheric boundary layer (ABL). The effects of the surface – heating, cooling, and friction – cause turbulent mixing within the air layer. Significant movement of heat, matter, or momentum on time scales of less than a day are caused by turbulent motions. Boundary layer meteorology includes the study of all types of surface–atmosphere boundary, including ocean, lake, urban land and non-urban land for the study of meteorology.

Dynamic Meteorology

Dynamic meteorology generally focuses on the fluid dynamics of the atmosphere. The idea of air parcel is used to define the smallest element of the atmosphere, while ignoring the discrete molecular and chemical nature of the atmosphere. An air parcel is defined as a point in the fluid continuum of the atmosphere. The fundamental laws of fluid dynamics, thermodynamics, and motion are used to study the atmosphere. The physical quantities that characterize the state of the atmosphere are temperature, density, pressure, etc. These variables have unique values in the continuum.

Applications

Weather Forecasting

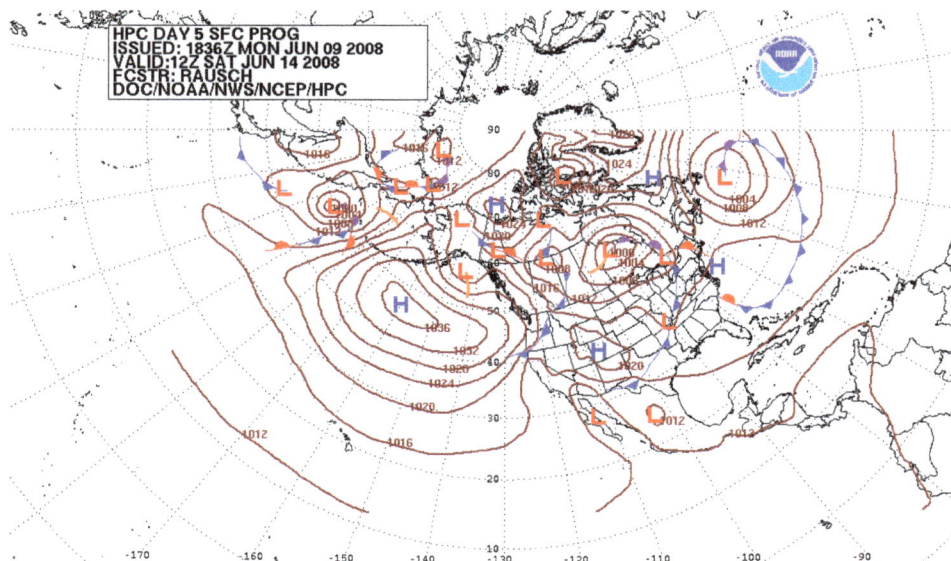

Forecast of surface pressures five days into the future for the north Pacific, North America, and north Atlantic Ocean

Weather forecasting is the application of science and technology to predict the state of the atmosphere at a future time and given location. Humans have attempted to predict the weather informally for millennia and formally since at least the 19th century. Weather forecasts are made by collecting quantitative data about the current state of the atmosphere and using scientific understanding of atmospheric processes to project how the atmosphere will evolve.

Once an all-human endeavor based mainly upon changes in barometric pressure, current weather conditions, and sky condition, forecast models are now used to determine future conditions. Human input is still required to pick the best possible forecast model to base the forecast upon, which involves pattern recognition skills, teleconnections, knowledge of model performance, and knowledge of model biases. The chaotic nature of the atmosphere, the massive computational power required to solve the equations that describe the atmosphere, error involved in measuring the initial conditions, and an incomplete understanding of atmospheric processes mean that forecasts become less accurate as the difference in current time and the time for which the forecast is being made (the *range* of the forecast) increases. The use of ensembles and model consensus help narrow the error and pick the most likely outcome.

There are a variety of end uses to weather forecasts. Weather warnings are important forecasts because they are used to protect life and property. Forecasts based on temperature and precipitation are important to agriculture, and therefore to commodity traders within stock markets. Temperature forecasts are used by utility companies to estimate demand over coming days. On an everyday basis, people use weather forecasts to determine what to wear on a given day. Since outdoor activities are severely curtailed by heavy rain, snow and the wind chill, forecasts can be used to plan activities around these events, and to plan ahead and survive them.

Aviation Meteorology

Aviation meteorology deals with the impact of weather on air traffic management. It is important for air crews to understand the implications of weather on their flight plan as well as their aircraft, as noted by the Aeronautical Information Manual:

The effects of ice on aircraft are cumulative—thrust is reduced, drag increases, lift lessens, and weight increases. The results are an increase in stall speed and a deterioration of aircraft performance. In extreme cases, 2 to 3 inches of ice can form on the leading edge of the airfoil in less than 5 minutes. It takes but 1/2 inch of ice to reduce the lifting power of some aircraft by 50 percent and increases the frictional drag by an equal percentage.

Agricultural Meteorology

Meteorologists, soil scientists, agricultural hydrologists, and agronomists are persons concerned with studying the effects of weather and climate on plant distribution, crop yield, water-use efficiency, phenology of plant and animal development, and the energy balance of managed and natural ecosystems. Conversely, they are interested in the role of vegetation on climate and weather.

Hydrometeorology

Hydrometeorology is the branch of meteorology that deals with the hydrologic cycle, the water budget, and the rainfall statistics of storms. A hydrometeorologist prepares and issues forecasts of accumulating (quantitative) precipitation, heavy rain, heavy snow, and highlights areas with the potential for flash flooding. Typically the range of knowledge that is required overlaps with climatology, mesoscale and synoptic meteorology, and other geosciences.

The multidisciplinary nature of the branch can result in technical challenges, since tools and solutions from each of the individual disciplines involved may behave slightly differently, be optimized for different hard- and software platforms and use different data formats. There are some initiatives - such as the DRIHM project - that are trying to address this issue.

Nuclear Meteorology

Nuclear meteorology investigates the distribution of radioactive aerosols and gases in the atmosphere.

Maritime Meteorology

Maritime meteorology deals with air and wave forecasts for ships operating at sea. Organizations such as the Ocean Prediction Center, Honolulu National Weather Service forecast office, United Kingdom Met Office, and JMA prepare high seas forecasts for the world's oceans.

Military Meteorology

Military meteorology is the research and application of meteorology for military purposes. In the United States, the United States Navy's Commander, Naval Meteorology and Oceanography Command oversees meteorological efforts for the Navy and Marine Corps while the United States Air Force's Air Force Weather Agency is responsible for the Air Force and Army.

Environmental Meteorology

Environmental meteorology mainly analyzes industrial pollution dispersion physically and chemically based on meteorological parameters such as temperature, humidity, wind, and various weather conditions.

Renewable Energy

Meteorology applications in renewable energy includes basic research, "exploration", and potential mapping of wind power and solar radiation for wind and solar energy.

Ecology

Ecology is the scientific analysis and study of interactions among organisms and their environment. It is an interdisciplinary field that includes biology, geography, and Earth science. Ecology includes the study of interactions organisms have with each other, other organisms, and with abiotic components of their environment. Topics of interest to ecologists include the diversity, distribution, amount (biomass), and number (population) of particular organisms, as well as cooperation and competition between organisms, both within and among ecosystems. Ecosystems are composed of dynamically interacting parts including organisms, the communities they make up, and the non-living components of their environment. Ecosystem processes, such as primary production, pedogenesis, nutrient cycling, and various niche construction activities, regulate the flux of energy and matter through an environment. These processes are sustained by organisms with specific life history traits, and the variety of organisms is called biodiversity. Biodiversity, which refers to the varieties of species, genes, and ecosystems, enhances certain ecosystem services.

Ecology is not synonymous with environment, environmentalism, natural history, or environmental science. It is closely related to evolutionary biology, genetics, and ethology. An important focus for ecologists is to improve the understanding of how biodiversity affects ecological function. Ecologists seek to explain:

- Life processes, interactions, and adaptations

- The movement of materials and energy through living communities

- The successional development of ecosystems

- The abundance and distribution of organisms and biodiversity in the context of the environment.

Ecology is a human science as well. There are many practical applications of ecology in conservation biology, wetland management, natural resource management (agroecology, agriculture, forestry, agroforestry, fisheries), city planning (urban ecology), community health, economics, basic and applied science, and human social interaction (human ecology). For example, the *Circles of Sustainability approach treats ecology as more than the environment ‹out there›. It is not treated as separate from humans. Organisms (including humans) and* resources compose ecosystems which, in turn, maintain biophysical feedback mechanisms that moderate processes acting on living (biotic) and non-living (abiotic) components of the planet. Ecosystems sustain life-supporting functions and produce natural capital like biomass production (food, fuel, fiber, and medicine), the regulation of climate, global biogeochemical cycles, water filtration, soil formation, erosion control, flood protection, and many other natural features of scientific, historical, economic, or intrinsic value.

The word «ecology» («Ökologie») was coined in 1866 by the German scientist Ernst Haeckel (1834–1919). Ecological thought is derivative of established currents in philosophy, particularly from ethics and politics. Ancient Greek philosophers such as Hippocrates and Aristotle laid the foundations of ecology in their studies on natural history. Modern ecology became a much more rigorous science in the late 19th century. Evolutionary concepts relating to adaptation and natural selection became the cornerstones of modern ecological theory.

Integrative Levels, Scope, and Scale of Organization

The scope of ecology contains a wide array of interacting levels of organization spanning micro-level (e.g., cells) to a planetary scale (e.g., biosphere) phenomena. Ecosystems, for example, contain abiotic resources and interacting life forms (i.e., individual organisms that aggregate into populations which aggregate into distinct ecological communities). Ecosystems are dynamic, they do not always follow a linear successional path, but they are always changing, sometimes rapidly and sometimes so slowly that it can take thousands of years for ecological processes to bring about certain successional stages of a forest. An ecosystem's area can vary greatly, from tiny to vast. A single tree is of little consequence to the classification of a forest ecosystem, but critically relevant to organisms living in and on it. Several generations of an aphid population can exist over the lifespan of a single leaf. Each of those aphids, in turn, support diverse bacterial communities. The nature of connections in ecological communities cannot be explained by knowing the details of each species in isolation, because the emergent pattern is neither revealed nor predicted until the ecosystem is studied as an integrated whole. Some ecological principles, however, do exhibit collective properties where the sum of the components explain the properties of the whole, such as birth rates of a population being equal to the sum of individual births over a designated time frame.

Hierarchical Ecology

System behaviors must first be arrayed into different levels of organization. Behaviors corresponding to higher levels occur at slow rates. Conversely, lower organizational levels exhibit rapid rates. For example, individual tree leaves respond rapidly to momentary changes in light intensity, CO_2 concentration, and the like. The growth of the tree responds more slowly and integrates these short-term changes.

O'Neill et al. (1986):76

The scale of ecological dynamics can operate like a closed system, such as aphids migrating on a single tree, while at the same time remain open with regard to broader scale influences, such as atmosphere or climate. Hence, ecologists classify ecosystems hierarchically by analyzing data collected from finer scale units, such as vegetation associations, climate, and soil types, and integrate this information to identify emergent patterns of uniform organization and processes that operate on local to regional, landscape, and chronological scales.

To structure the study of ecology into a conceptually manageable framework, the biological world is organized into a nested hierarchy, ranging in scale from genes, to cells, to tissues, to organs, to organisms, to species, to populations, to communities, to ecosystems, to biomes, and up to the level of the biosphere. This framework forms a panarchy and exhibits non-linear behaviors; this means that "effect and cause are disproportionate, so that small changes to critical variables, such as the number of nitrogen fixers, can lead to disproportionate, perhaps irreversible, changes in the system properties."

Biodiversity

Biodiversity refers to the variety of life and its processes. It includes the variety of living organisms, the genetic differences among them, the communities and ecosystems in which they occur, and the ecological and evolutionary processes that keep them functioning, yet ever changing and adapting.

Noss & Carpenter (1994):5

Biodiversity (an abbreviation of "biological diversity") describes the diversity of life from genes to ecosystems and spans every level of biological organization. The term has several interpretations, and there are many ways to index, measure, characterize, and represent its complex organization. Biodiversity includes species diversity, ecosystem diversity, and genetic diversity and scientists are interested in the way that this diversity affects the complex ecological processes operating at and among these respective levels. Biodiversity plays an important role in ecosystem services which by definition maintain and improve human quality of life. Preventing species extinctions is one way to preserve biodiversity and that goal rests on techniques that preserve genetic diversity, habitat and the ability for species to migrate. Conservation priorities and management techniques require different approaches and considerations to address the full ecological scope of biodiversity. Natural capital that supports populations is critical for maintaining ecosystem services and species migration (e.g., riverine fish runs and avian insect control) has been implicated as one mechanism by which those service losses are experienced. An understanding of biodiversity has practical applications for species and ecosystem-level conservation planners as they make management recommendations to consulting firms, governments, and industry.

Habitat

The habitat of a species describes the environment over which a species is known to occur and the type of community that is formed as a result. More specifically, "habitats can be defined as regions in environmental space that are composed of multiple dimensions, each representing a biotic or abiotic environmental variable; that is, any component or characteristic of the environment related directly (e.g. forage biomass and quality) or indirectly (e.g. elevation) to the use of a location by the animal.":745 For example, a habitat might be an aquatic or terrestrial environment that can be further categorized as a montane or alpine ecosystem. Habitat shifts provide important evidence of competition in nature where one population changes relative to the habitats that most other individuals of the species occupy. For example, one population of a species of tropical lizards (*Tropidurus hispidus*) has a flattened body relative to the main populations that live in open savanna. The population that lives in an isolated rock outcrop hides in crevasses where its flattened body offers a selective advantage. Habitat shifts also occur in the developmental life history of amphibians, and in insects that transition from aquatic to terrestrial habitats. Biotope and habitat are sometimes used interchangeably, but the former applies to a community's environment, whereas the latter applies to a species' environment.

Additionally, some species are ecosystem engineers, altering the environment within a localized region. For instance, beavers manage water levels by building dams which improves their habitat in a landscape.

Niche

Definitions of the niche date back to 1917, but G. Evelyn Hutchinson made conceptual advances in 1957 by introducing a widely adopted definition: "the set of biotic and abiotic conditions in which a species is able to persist and maintain stable population sizes." The ecological niche is a cen-tral concept in the ecology of organisms and is sub-divided into the *fundamental* and the *realized* niche. The fundamental niche is the set of environmental conditions under which a species is able to persist. The realized niche is the set of environmental plus ecological conditions under which a species persists. The Hutchinsonian niche is defined more technically as a "Euclidean hyperspace whose *dimensions* are defined as environmental variables and whose size is a function of the number of values that the environmental values may assume for which an organism has *positive fitness."*

Termite mounds with varied heights of chimneys regulate gas exchange, temperature and other environmental parameters that are needed to sustain the internal physiology of the entire colony.

Biogeographical patterns and range distributions are explained or predicted through knowledge of a species' traits and niche requirements. Species have functional traits that are uniquely adapted to the ecological niche. A trait is a measurable property, phenotype, or characteristic of an organism that may influence its survival. Genes play an important role in the interplay of development and environmental expression of traits. Resident species evolve traits that are fitted to the selection pressures of their local environment. This tends to afford them a competitive advantage and discourages similarly adapted species from having an overlapping geographic range. The competitive exclusion principle states that two species cannot coexist indefinitely by living off the same limiting resource; one will always out-compete the other. When similarly adapted species overlap geographically, closer inspection reveals subtle ecological differences in their habitat or dietary requirements. Some models and empirical studies, however, suggest that disturbances can stabilize the co-evolution and shared niche occupancy of similar species inhabiting species-rich communities. The habitat plus the niche is called the ecotope, which is defined as the full range of environmental and biological variables affecting an entire species.

Niche Construction

Organisms are subject to environmental pressures, but they also modify their habitats. The regulatory feedback between organisms and their environment can affect conditions from local (e.g., a beaver pond) to global scales, over time and even after death, such as decaying logs or silica skeleton deposits from marine organisms. The process and concept of ecosystem engineering is related to niche construction, but the former relates only to the physical modifications of the habitat whereas the latter also considers the evolutionary implications of physical changes to the environment and the feedback this causes on the process of natural selection. Ecosystem engineers are defined as: "organisms that directly or indirectly modulate the availability of resources to other species, by causing physical state changes in biotic or abiotic materials. In so doing they modify, maintain and create habitats."

The ecosystem engineering concept has stimulated a new appreciation for the influence that organisms have on the ecosystem and evolutionary process. The term "niche construction" is more often used in reference to the under-appreciated feedback mechanisms of natural selection imparting forces on the abiotic niche. An example of natural selection through ecosystem engineering occurs in the nests of social insects, including ants, bees, wasps, and termites. There is an emergent homeostasis or homeorhesis in the structure of the nest that regulates, maintains and defends the physiology of the entire colony. Termite mounds, for example, maintain a constant internal temperature through the design of air-conditioning chimneys. The structure of the nests themselves are subject to the forces of natural selection. Moreover, a nest can survive over successive generations, so that progeny inherit both genetic material and a legacy niche that was constructed before their time.

Biome

Biomes are larger units of organization that categorize regions of the Earth's ecosystems, mainly according to the structure and composition of vegetation. There are different methods to define the continental boundaries of biomes dominated by different functional types of vegetative communities that are limited in distribution by climate, precipitation, weather and other environmental variables. Biomes include tropical rainforest, temperate broadleaf and mixed forest, temperate deciduous forest, taiga, tundra, hot desert, and polar desert. Other researchers have recently categorized other biomes, such as the human and oceanic microbiomes. To a microbe, the human body is a habitat and a landscape. Microbiomes were discovered largely through advances in molecular genetics, which have revealed a hidden richness of microbial diversity on the planet. The oceanic microbiome plays a significant role in the ecological biogeochemistry of the planet's oceans.

Biosphere

The largest scale of ecological organization is the biosphere: the total sum of ecosystems on the planet. Ecological relationships regulate the flux of energy, nutrients, and climate all the way up to the planetary scale. For example, the dynamic history of the planetary atmosphere's CO_2 and O_2 composition has been affected by the biogenic flux of gases coming from respiration and photosynthesis, with levels fluctuating over time in relation to the ecology and evolution of plants and animals. Ecological theory has also been used to explain self-emergent regulatory phenomena at the planetary scale: for example, the Gaia hypothesis is an example of holism applied in ecological theory. The Gaia hypothesis states that there is an emergent feedback loop generated by the metabolism of living organisms that maintains the core temperature of the Earth and atmospheric conditions within a narrow self-regulating range of tolerance.

Individual Ecology

Understanding traits of individual organisms helps explain patterns and processes at other levels of organization including populations, communities, and ecosystems. Several areas of ecology of evolution that focus on such traits are life history theory, ecophysiology, metabolic theory of ecology, and Ethology. Examples of such traits include features of an organisms life cycle such as age to maturity, life span, or metabolic costs of reproduction. Other traits may be related to structure, such as the spines of a cactus or dorsal spines of a bluegill sunfish, or behaviors such as courtship displays or pair bonding. Other traits include emergent properties that are the result at least in part of interactions with the surrounding environment such as growth rate, resource uptake rate, winter, and deciduous vs. drought deciduous trees and shrubs.

One set of characteristics relate to body size and temperature. The metabolic theory of ecology provides a predictive qualitative set of relationships between an organism's body size and temperature and metabolic processes. In general, smaller, warmer organisms have higher metabolic rates and this results in a variety of predictions regarding individual somatic growth rates, reproduction and population growth rates, population size, and resource uptake rates.

The traits of organisms are subject to change through acclimation, development, and evolution. For this reason, individuals form a shared focus for ecology and for evolutionary ecology.

Population Ecology

Population ecology studies the dynamics of specie populations and how these populations interact with the wider environment. A population consists of individuals of the same species that live, interact, and migrate through the same niche and habitat.

A primary law of population ecology is the Malthusian growth model which states, "a population will grow (or decline) exponentially as long as the environment experienced by all individuals in the population remains constant." Simplified population models usually start with four variables: death, birth, immigration, and emigration.

An example of an introductory population model describes a closed population, such as on an island, where immigration and emigration does not take place. Hypotheses are evaluated with reference to a null hypothesis which states that random processes create the observed data. In these island models, the rate of population change is described by:

$$\frac{dN}{dT} = bN - dN = (b - d)N = rN,$$

where N is the total number of individuals in the population, b and d are the per capita rates of birth and death respectively, and r is the per capita rate of population change.

Using these modelling techniques, Malthus' population principle of growth was later transformed into a model known as the logistic equation:

$$\frac{dN}{dT} = aN\left(1 - \frac{N}{K}\right),$$

where N is the number of individuals measured as biomass density, a is the maximum per-capita rate of change, and K is the carrying capacity of the population. The formula states that the rate of change in population size (dN/dT) is equal to growth (aN) that is limited by carrying capacity ($1 - N/K$).

Population ecology builds upon these introductory models to further understand demographic processes in real study populations. Commonly used types of data include life history, fecundity, and survivorship, and these are analysed using mathematical techniques such as matrix algebra. The information is used for managing wildlife stocks and setting harvest quotas. In cases where basic models are insufficient, ecologists may adopt different kinds of statistical methods, such as the Akaike information criterion, or use models that can become mathematically complex as "several competing hypotheses are simultaneously confronted with the data."

Metapopulations and Migration

The concept of metapopulations was defined in 1969 as "a population of populations which go extinct locally and recolonize". Metapopulation ecology is another statistical approach that is often used in conservation research. Metapopulation models simplify the landscape into patches of varying lev-els of quality, and metapopulations are linked by the migratory behaviours of organisms. Animal migration is set apart from other kinds of movement; because, it involves the seasonal departure and return of individuals from a habitat. Migration is also a population-level phenomenon, as with the migration routes followed by plants as they occupied northern post-glacial environments. Plant ecologists use pollen records that accumulate and stratify in wetlands to reconstruct the timing of plant migration and dispersal relative to historic and contemporary climates. These migration routes involved an expansion of the range as plant populations expanded from one area to another. There is a larger taxonomy of movement, such as commuting, foraging, territorial behaviour, stasis, and ranging. Dispersal is usually distinguished from migration; because, it involves the one way perma-nent movement of individuals from their birth population into another population.

In metapopulation terminology, migrating individuals are classed as emigrants (when they leave a region) or immigrants (when they enter a region), and sites are classed either as sources or sinks. A site is a generic term that refers to places where ecologists sample populations, such as ponds or defined sampling areas in a forest. Source patches are productive sites that generate a seasonal supply of juveniles that migrate to other patch locations. Sink patches are unproductive sites that only receive migrants; the population at the site will disappear unless rescued by an adjacent source patch or environmental conditions become more favourable. Metapopulation models examine patch dynamics over time to answer potential questions about spatial and demographic ecology. The ecology of metapopulations is a dynamic process of extinction and colonization. Small patches of lower quality (i.e., sinks) are maintained or rescued by a seasonal influx of new immigrants. A dynamic metapopulation structure evolves from year to year, where some patches are sinks in dry years and are sources when conditions are more favourable. Ecologists use a mixture of computer models and field studies to explain metapopulation structure.

Community Ecology

Community ecology examines how interactions among species and their environment affect the abundance, distribution and diversity of species within communities.

Johnson & Stinchcomb (2007):250

Community ecology is the study of the interactions among a collections of species that inhabit the same geographic area. Community ecologists study the determinants of patterns and processes for two or more interacting species. Research in community ecology might measure species diversity in grasslands in relation to soil fertility. It might also include the analysis of predator-prey dynamics, competition among similar plant species, or mutualistic interactions between crabs and corals

Interspecific interactions such as predation are a key aspect of community ecology.

Ecosystem Ecology

These ecosystems, as we may call them, are of the most various kinds and sizes. They form one category of the multitudinous physical systems of the universe, which range from the universe as a whole down to the atom.

Tansley (1935):299

Ecosystems may be habitats within biomes that form an integrated whole and a dynamically responsive system having both physical and biological complexes. Ecosystem ecology is the science of determining the fluxes of materials (e.g. carbon, phosphorus) between different pools (e.g., tree biomass, soil organic material). Ecosystem ecologist attempt to determine the underlying causes of these fluxes. Research in ecosystem ecology might measure primary production (g C/m^2) in a wetland in relation to decomposition and consumption rates (g $C/m^2/y$). This requires an understanding of the community connections between plants (i.e., primary producers) and the decomposers (e.g., fungi and bacteria),

The underlying concept of ecosystem can be traced back to 1864 in the published work of George Perkins Marsh ("Man and Nature"). Within an ecosystem, organisms are linked to the physical and biological components of their environment to which they are adapted. Ecosystems are complex adaptive systems where the interaction of life processes form self-organizing patterns across different scales of time and space. Ecosystems are broadly categorized as terrestrial, freshwater, atmospheric, or marine. Differences stem from the nature of the unique physical environments that shapes the biodiversity within each. A more recent addition to ecosystem ecology are technoecosystems, which are affected by or primarily the result of human activity.

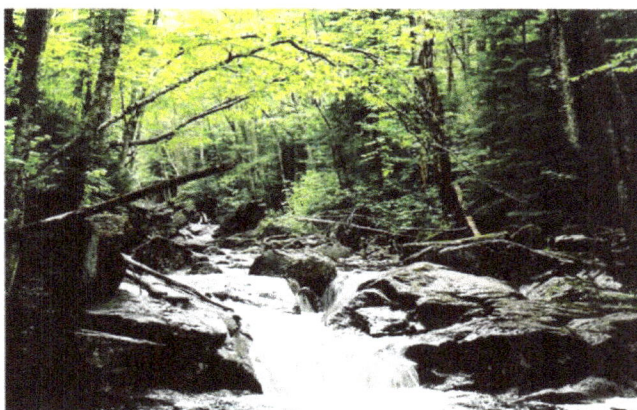

A riparian forest in the White Mountains, New Hampshire (USA), an example of ecosystem ecology

Food Webs

A food web is the archetypal ecological network. Plants capture solar energy and use it to synthesize simple sugars during photosynthesis. As plants grow, they accumulate nutrients and are eaten by grazing herbivores, and the energy is transferred through a chain of organisms by consumption. The simplified linear feeding pathways that move from a basal trophic species to a top consumer is called the food chain. The larger interlocking pattern of food chains in an ecological community creates a complex food web. Food webs are a type of concept map or a heuristic device that is used to illustrate and study pathways of energy and material flows.

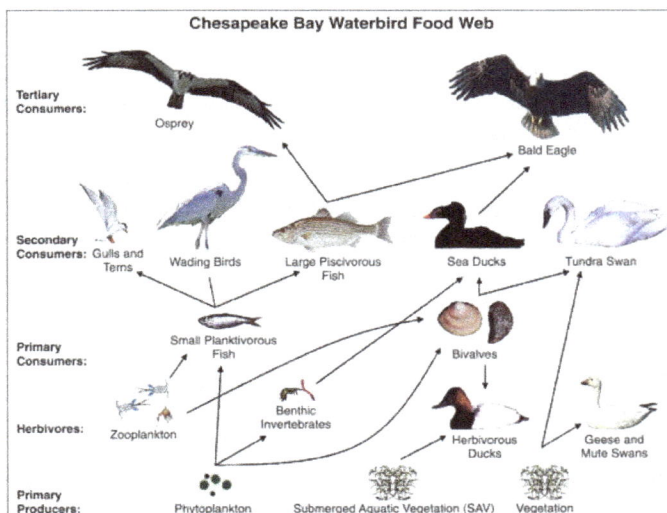

Generalized food web of waterbirds from Chesapeake Bay

Food webs are often limited relative to the real world. Complete empirical measurements are generally restricted to a specific habitat, such as a cave or a pond, and principles gleaned from food web microcosm studies are extrapolated to larger systems. Feeding relations require extensive investigations into the gut contents of organisms, which can be difficult to decipher, or stable isotopes can be used to trace the flow of nutrient diets and energy through a food web. Despite these limitations, food webs remain a valuable tool in understanding community ecosystems.

Food webs exhibit principles of ecological emergence through the nature of trophic relationships: some species have many weak feeding links (e.g., omnivores) while some are more specialized with fewer stronger feeding links (e.g., primary predators). Theoretical and empirical studies identify non-random emergent patterns of few strong and many weak linkages that explain how ecological communities remain stable over time. Food webs are composed of subgroups where members in a community are linked by strong interactions, and the weak interactions occur between these subgroups. This increases food web stability. Step by step lines or relations are drawn until a web of life is illustrated.

Trophic Levels

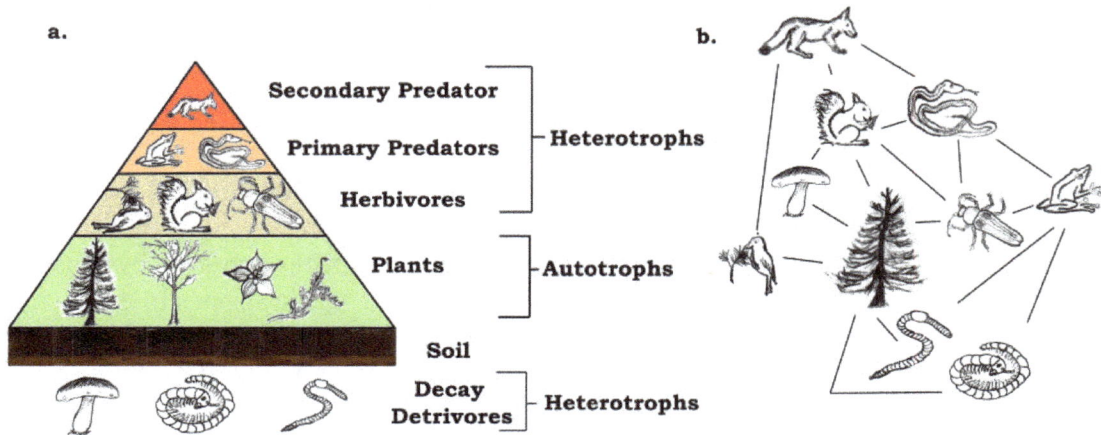

A trophic pyramid (a) and a food-web (b) illustrating ecological relationships among creatures that are typical of a northern boreal terrestrial ecosystem. The trophic pyramid roughly represents the biomass (usually measured as total dry-weight) at each level. Plants generally have the greatest biomass. Names of trophic categories are shown to the right of the pyramid. Some ecosystems, such as many wetlands, do not organize as a strict pyramid, because aquatic plants are not as productive as long-lived terrestrial plants such as trees. Ecological trophic pyramids are typically one of three kinds: 1) pyramid of numbers, 2) pyramid of biomass, or 3) pyramid of energy.

A trophic level is "a group of or-ganisms acquiring a considerable majority of its energy from the adjacent level nearer the abiotic source." Links in food webs primarily connect feeding relations or trophism among species. Biodiversity within ecosystems can be organized into trophic pyramids, in which the vertical dimension represents feeding relations that become further removed from the base of the food chain up toward top predators, and the horizontal dimension represents the abundance or biomass at each level. When the relative abundance or biomass of each species is sorted into its respective trophic level, they naturally sort into a 'pyramid of numbers'.

Species are broadly categorized as autotrophs (or primary producers), heterotrophs (or consumers), and Detritivores (or decomposers). Autotrophs are organisms that produce their own food (production is greater than respiration) by photosynthesis or chemosynthesis. Heterotrophs are organisms that must feed on others for nourishment and energy (respiration exceeds production). Heterotrophs can be further sub-divided into different functional groups, including primary consumers (strict herbivores), secondary consumers (carnivorous predators that feed exclusively on herbivores), and tertiary consumers (predators that feed on a mix of herbivores and predators). Omnivores do not fit neatly into a functional category because they eat both plant and animal tissues. It has been suggested that omnivores have a greater functional influence as predators, because compared to herbivores, they are relatively inefficient at grazing.

Trophic levels are part of the holistic or complex systems view of ecosystems. Each trophic level contains unrelated species that are grouped together because they share common ecological functions, giving a macroscopic view of the system. While the notion of trophic levels provides insight into energy flow and top-down control within food webs, it is troubled by the prevalence of omnivory in real ecosystems. This has led some ecologists to "reiterate that the notion that species clearly aggregate into discrete, homogeneous trophic levels is fiction." Nonetheless, recent studies have shown that real trophic levels do exist, but "above the herbivore trophic level, food webs are better characterized as a tangled web of omnivores."

Keystone Species

Sea otters, an example of a keystone species

A keystone species is a species that is connected to a disproportionately large number of other species in the food-web. Keystone species have lower levels of biomass in the trophic pyramid relative to the importance of their role. The many connections that a keystone species holds means that it maintains the organization and structure of entire communities. The loss of a keystone species results in a range of dramatic cascading effects that alters trophic dynamics, other food web connections, and can cause the extinction of other species.

Sea otters (*Enhydra lutris)* are commonly cited as an example of a keystone species; because, they limit the density of sea urchins that feed on kelp. If sea otters are removed from the system, the urchins graze until the kelp beds disappear, and this has a dramatic effect on community structure. Hunting of sea otters, for example, is thought to have led indirectly to the extinction of the Steller's sea cow (*Hydrodamalis gigas*). While the keystone species concept has been used extensively as a conservation tool, it has been criticized for being poorly defined from an operational stance. It is difficult to experimentally determine what species may hold a keystone role in each ecosystem. Furthermore, food web theory suggests that keystone species may not be common, so it is unclear how generally the keystone species model can be applied.

Ecological Complexity

Complexity is understood as a large computational effort needed to piece together numerous interacting parts exceeding the iterative memory capacity of the human mind. Global patterns of biological diversity are complex. This biocomplexity stems from the interplay among ecological processes that operate and influence patterns at different scales that grade into each other, such as transitional areas or ecotones spanning landscapes. Complexity stems from the interplay among levels of biological organization as energy, and matter is integrated into larger units that superimpose onto the smaller parts. "What were wholes on one level become parts on a higher one." Small scale patterns do not necessarily explain large scale phenomena, otherwise captured in the expression (coined by Aristotle) 'the sum is greater than the parts'.

"Complexity in ecology is of at least six distinct types: spatial, temporal, structural, process, behavioral, and geometric." From these principles, ecologists have identified emergent and self-organizing phenomena that operate at different environmental scales of influence, ranging from molecular to planetary, and these require different explanations at each integrative level. Ecological complexity relates to the dynamic resilience of ecosystems that transition to multiple shifting steady-states directed by random fluctuations of history. Long-term ecological studies provide important track records to better understand the complexity and resilience of ecosystems over longer temporal and broader spatial scales. These studies are managed by the International Long Term Ecological Network (LTER). The longest experiment in existence is the Park Grass Experiment, which was initiated in 1856. Another example is the Hubbard Brook study, which has been in operation since 1960.

Holism

Holism remains a critical part of the theoretical foundation in contemporary ecological studies. Holism addresses the biological organization of life that self-organizes into layers of emergent whole systems that function according to non-reducible properties. This means that higher order patterns of a whole functional system, such as an ecosystem, cannot be predicted or understood by a simple summation of the parts. "New properties emerge because the components interact, not because the basic nature of the components is changed."

Ecological studies are necessarily holistic as opposed to reductionistic. Holism has three scientific meanings or uses that identify with ecology: 1) the mechanistic complexity of ecosystems, 2) the practical description of patterns in quantitative reductionist terms where correlations may be identified but nothing is understood about the causal relations without reference to the whole system, which leads to 3) a metaphysical hierarchy whereby the causal relations of larger systems are understood without reference to the smaller parts. Scientific holism differs from mysticism that has appropriated the same term. An example of metaphysical holism is identified in the trend of increased exterior thickness in shells of different species. The reason for a thickness increase can be understood through reference to principles of natural selection via predation without need to reference or understand the biomolecular properties of the exterior shells.

Relation to Evolution

Ecology and evolution are considered sister disciplines of the life sciences. Natural selection, life history, development, adaptation, populations, and inheritance are examples of concepts that thread equally into ecological and evolutionary theory. Morphological, behavioural, and genetic traits, for example, can be mapped onto evolutionary trees to study the historical development of a species in relation to their functions and roles in different ecological circumstances. In this framework, the analytical tools of ecologists and evolutionists overlap as they organize, classify, and investigate life through common systematic principals, such as phylogenetics or the Linnaean system of taxonomy. The two disciplines often appear together, such as in the title of the journal Trends in Ecology and Evolution. There is no sharp boundary separating ecology from evolution, and they differ more in their areas of applied focus. Both disciplines discover and explain emergent and unique properties and processes operating across different spatial or temporal scales of organization. While the boundary between ecology and evolution is not always clear, ecologists study the abiotic and biotic factors that influence evolutionary processes, and evolution can be rapid, occurring on ecological timescales as short as one generation.

Behavioural Ecology

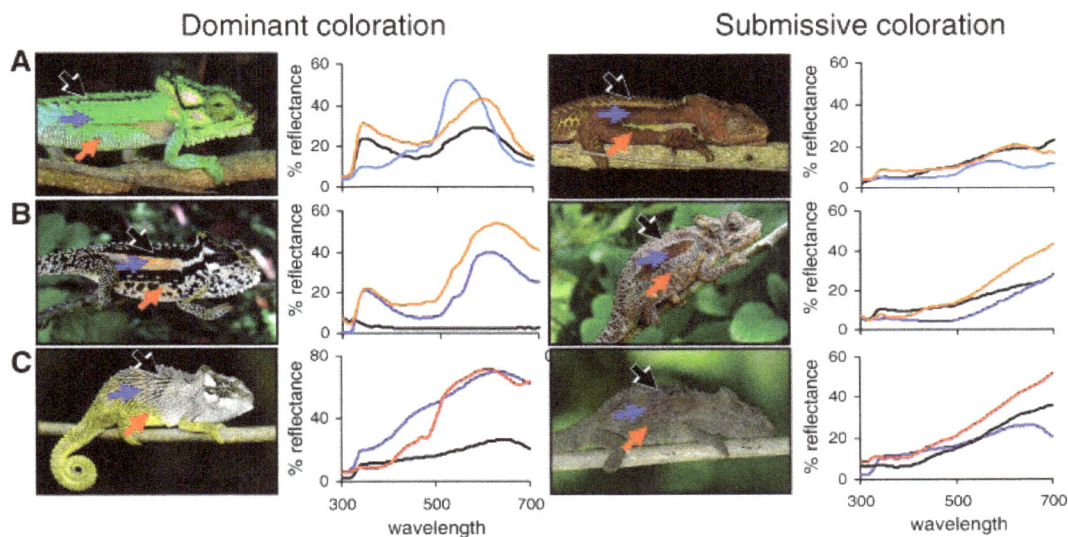

Social display and colour variation in differently adapted species of chameleons (Bradypodion spp.). Chameleons change their skin colour to match their background as a behavioural defence mechanism and also use colour to communicate with other members of their species, such as dominant (left) versus submissive (right) patterns shown in the three species (A-C) above.

All organisms can exhibit behaviours. Even plants express complex behaviour, including memory and communication. Behavioural ecology is the study of an organism's behaviour in its environment and its ecological and evolutionary implications. Ethology is the study of observable movement or behaviour in animals. This could include investigations of motile sperm of plants, mobile phytoplankton, zooplankton swimming toward the female egg, the cultivation of fungi by weevils, the mating dance of a salamander, or social gatherings of amoeba.

Adaptation is the central unifying concept in behavioural ecology. Behaviours can be recorded as traits and inherited in much the same way that eye and hair colour can. Behaviours can evolve by means of natural selection as adaptive traits conferring functional utilities that increases reproductive fitness.

Predator-prey interactions are an introductory concept into food-web studies as well as behavioural ecology. Prey species can exhibit different kinds of behavioural adaptations to predators, such as avoid, flee, or defend. Many prey species are faced with multiple predators that differ in the degree of danger posed. To be adapted to their environment and face predatory threats, organisms must balance their energy budgets as they invest in different aspects of their life history, such as growth, feeding, mating, socializing, or modifying their habitat. Hypotheses posited in behavioural ecology are generally based on adaptive principles of conservation, optimization, or efficiency. For example, "[t]he threat-sensitive predator avoidance hypothesis predicts that prey should assess the degree of threat posed by different predators and match their behaviour according to current levels of risk" or "[t]he optimal flight initiation distance occurs where expected postencounter fitness is maximized, which depends on the prey's initial fitness, benefits obtainable by not fleeing, energetic escape costs, and expected fitness loss due to predation risk."

Symbiosis: Leafhoppers (*Eurymela fenestrata)* are protected by ants (*Iridomyrmex purpureus*) in a symbiotic relationship. The ants protect the leafhoppers from predators and in return the leafhoppers feeding on plants exude honeydew from their anus that provides energy and nutrients to tending ants.

Elaborate sexual displays and posturing are encountered in the behavioural ecology of animals. The birds-of-paradise, for example, sing and display elaborate ornaments during courtship. These displays serve a dual purpose of signalling healthy or well-adapted individuals and desirable genes. The displays are driven by sexual selection as an advertisement of quality of traits among suitors.

Cognitive Ecology

Cognitive ecology integrates theory and observations from evolutionary ecology and neurobiology, primarily cognitive science, in order to understand the effect that animal interaction with their habitat has on their cognitive systems and how those systems restrict behavior within an ecological and evolutionary framework. "Until recently, however, cognitive scientists have not paid sufficient attention to the fundamental fact that cognitive traits evolved under particular natural settings. With consideration of the selection pressure on cognition, cognitive ecology can contribute intellectual coherence to the multidisciplinary study of cognition." As a study involving the 'coupling' or interactions between organism and environment, cognitive ecology is closely related to enactivism, a field based upon the view that "...we must see the organism and environment as bound together in reciprocal specification and selection...".

Social Ecology

Social ecological behaviours are notable in the social insects, slime moulds, social spiders, human society, and naked mole-rats where eusocialism has evolved. Social behaviours include reciprocally beneficial behaviours among kin and nest mates and evolve from kin and group selection. Kin selection explains altruism through genetic relationships, whereby an altruistic behaviour leading to death is rewarded by the survival of genetic copies distributed among surviving relatives. The social insects, including ants, bees, and wasps are most famously studied for this type of relationship because the male drones are clones that share the same genetic make-up as every other male in the colony. In contrast, group selectionists find examples of altruism among non-genetic relatives and explain this through selection acting on the group; whereby, it becomes selectively advantageous for groups if their members express altruistic behaviours to one another. Groups with predominantly altruistic members beat groups with predominantly selfish members.

Coevolution

Bumblebees and the flowers they pollinate have coevolved so that both have become dependent on each other for survival.

Ecological interactions can be classified broadly into a host and an associate relationship. A host is any entity that harbours another that is called the associate. Relationships within a species that are mutually or reciprocally beneficial are called mutualisms. Examples of mutualism include fungus-growing ants employing agricultural symbiosis, bacteria living in the guts of insects and other organisms, the fig wasp and yucca moth pollination complex, lichens with fungi and photosynthetic algae, and corals with photosynthetic algae. If there is a physical connection between host and associate, the relationship is called symbiosis. Approximately 60% of all plants, for example, have a symbiotic relationship with arbuscular mycorrhizal fungi living in their roots forming an exchange network of carbohydrates for mineral nutrients.

Indirect mutualisms occur where the organisms live apart. For example, trees living in the equatorial regions of the planet supply oxygen into the atmosphere that sustains species living in distant polar regions of the planet. This relationship is called commensalism; because, many others receive the benefits of clean air at no cost or harm to trees supplying the oxygen. If the associate benefits while the host suffers, the relationship is called parasitism. Although parasites impose a cost to their host (e.g., via damage to their reproductive organs or propagules, denying the services of a beneficial partner), their net effect on host fitness is not necessarily negative and, thus, becomes difficult to forecast. Co-evolution is also driven by competition among species or among members of the same species under the banner of reciprocal antagonism, such as grasses competing for growth space. The Red Queen Hypothesis, for example, posits that parasites track down and specialize on the locally common genetic defense systems of its host that drives the evolution of sexual reproduction to diversify the genetic constituency of populations responding to the antagonistic pressure.

Parasitism: A harvestman arachnid being parasitized by mites. The harvestman is being consumed, while the mites benefit from traveling on and feeding off of their host.

Biogeography

Biogeography (an amalgamation of biology and geography) is the comparative study of the geographic distribution of organisms and the corresponding evolution of their traits in space and time. The *Journal of Biogeography* was established in 1974. Biogeography and ecology share many of their disciplinary roots. For example, the theory of island biogeography, published by the mathematician Robert MacArthur and ecologist Edward O. Wilson in 1967 is considered one of the fundamentals of ecological theory.

Biogeography has a long history in the natural sciences concerning the spatial distribution of plants and animals. Ecology and evolution provide the explanatory context for biogeographical studies. Biogeographical patterns result from ecological processes that influence range distributions, such as migration and dispersal. and from historical processes that split populations or species into different areas. The biogeographic processes that result in the natural splitting of species explains much of the modern distribution of the Earth's biota. The splitting of lineages in a species is called vicariance biogeography and it is a sub-discipline of biogeography. There are also practical applications in the field of biogeography concerning ecological systems and processes. For example, the range and distribution of biodiversity and invasive species responding to climate change is a serious concern and active area of research in the context of global warming.

R/K-Selection Theory

A population ecology concept is r/K selection theory,[D] one of the first predictive models in ecology used to explain life-history evolution. The premise behind the r/K selection model is that natural selection pressures change according to population density. For example, when an island is first colonized, density of individuals is low. The initial increase in population size is not limited by competition, leaving an abundance of available resources for rapid population growth. These early phases of population growth experience *density-independent* forces of natural selection, which is called r-selection. As the population becomes more crowded, it approaches the island's carrying capacity, thus forcing individuals to compete more heavily for fewer available resources. Under crowded conditions, the population experiences density-dependent forces of natural selection, called *K*-selection.

In the *r/K*-selection model, the first variable r is the intrinsic rate of natural increase in population size and the second variable *K* is the carrying capacity of a population. Different species evolve different life-history strategies spanning a continuum between these two selective forces. An *r*-selected species is one that has high birth rates, low levels of parental investment, and high rates of mortality before individuals reach maturity. Evolution favours high rates of fecundity in *r*-selected species. Many kinds of insects and invasive species exhibit *r*-selected characteristics. In contrast, a *K*-selected species has low rates of fecundity, high levels of parental investment in the young, and low rates of mortality as individuals mature. Humans and elephants are examples of species exhibiting K-selected characteristics, including longevity and efficiency in the conversion of more resources into fewer offspring.

Molecular Ecology

The important relationship between ecology and genetic inheritance predates modern techniques for molecular analysis. Molecular ecological research became more feasible with the development of rapid and accessible genetic technologies, such as the polymerase chain reaction (PCR). The rise of molecular technologies and influx of research questions into this new ecological field resulted in the publication *Molecular Ecology* in 1992. Molecular ecology uses various analytical techniques to study genes in an evolutionary and ecological context. In 1994, John Avise also played a leading role in this area of science with the publication of his book, Molecular Markers, Natural History and Evolution. Newer technologies opened a wave of genetic analysis into organisms once difficult to study from an ecological or evolutionary standpoint, such as bacteria, fungi, and nematodes. Molecular ecology engendered a new research paradigm for investigating ecological questions considered otherwise intractable. Molecular investigations revealed previously obscured details in the tiny intricacies of nature and improved resolution into probing questions about behavioural and biogeographical ecology. For example, molecular ecology revealed promiscuous sexual behaviour and multiple male partners in tree swallows previously thought to be socially monogamous. In a biogeographical context, the marriage between genetics, ecology, and evolution resulted in a new sub-discipline called phylogeography.

Human Ecology

The history of life on Earth has been a history of interaction between living things and their surroundings. To a large extent, the physical form and the habits of the earth's vegetation and its animal life have been molded by the environment. Considering the whole span of earthly time, the opposite effect, in which life actually modifies its surroundings, has been relatively slight. Only within the moment of time represented by the present century has one species man acquired significant power to alter the nature of his world.

Rachel Carson, "Silent Spring"

Ecology is as much a biological science as it is a human science. Human ecology is an interdisciplinary investigation into the ecology of our species. "Human ecology may be defined: (1) from a bio-ecological standpoint as the study of man as the ecological dominant in plant and animal communities and systems; (2) from a bio-ecological standpoint as simply another animal affecting and being affected by his physical environment; and (3) as a human being, somehow different from animal life in general, interacting with physical and modified environments in a distinctive and creative way. A truly interdisciplinary human ecology will most likely address itself to all three." The term was formally introduced in 1921, but many sociologists, geographers, psychologists, and other disciplines were interested in human relations to natural systems centuries prior, especially in the late 19th century.

The ecological complexities human beings are facing through the technological transformation of the planetary biome has brought on the Anthropocene. The unique set of circumstances has generated the need for a new unifying science called coupled human and natural systems that builds upon, but moves beyond the field of human ecology. Ecosystems tie into human societies through the critical and all encompassing life-supporting functions they sustain. In recognition of these functions and the incapability of traditional economic valuation methods to see the value in ecosystems, there has been a surge of interest in social-natural capital, which provides the means to put a value on the stock and use of information and materials stemming from ecosystem goods and services. Ecosystems produce, regulate, maintain, and supply services of critical necessity and beneficial to human health (cognitive and physiological), economies, and they even provide an information or reference function as a living library giving opportunities for science and cognitive development in children engaged in the complexity of the natural world. Ecosystems relate importantly to human ecology as they are the ultimate base foundation of global economics as every commodity, and the capacity for exchange ultimately stems from the ecosystems on Earth.

Restoration and Management

Ecosystem management is not just about science nor is it simply an extension of traditional resource management; it offers a fundamental reframing of how humans may work with nature.

Grumbine (1994):27

Ecology is an employed science of restoration, repairing disturbed sites through human intervention, in natural resource management, and in environmental impact assessments. Edward O. Wilson predicted in 1992 that the 21st century "will be the era of restoration in ecology". Ecological science has boomed in the industrial investment of restoring ecosystems and their processes in abandoned sites after disturbance. Natural resource managers, in forestry, for example, employ ecologists to develop, adapt, and implement ecosystem based methods into the planning, operation, and restoration phases of land-use. Ecological science is used in the methods of sustainable harvesting, disease, and fire outbreak management, in fisheries stock management, for integrating land-use with protected areas and communities, and conservation in complex geo-political landscapes.

Relation to the Environment

The environment of ecosystems includes both physical parameters and biotic attributes. It is dynamically interlinked, and contains resources for organisms at any time throughout their life cycle. Like "ecology", the term "environment" has different conceptual meanings and overlaps with the concept of "nature". Environment "includes the physical world, the social world of human relations and the built world of human creation." The physical environment is external to the level of biological organization under investigation, including abiotic factors such as tem-perature, radiation, light, chemistry, climate and geology. The biotic environment includes genes, cells, organisms, members of the same species (conspecifics) and other species that share a habitat.

The distinction between external and internal environments, however, is an abstraction parsing life and environment into units or facts that are inseparable in reality. There is an interpenetration of cause and effect between the environment and life. The laws of thermodynamics, for example, apply to ecology by means of its physical state. With an understanding of metabolic and thermodynamic principles, a complete accounting of energy and material flow can be traced through an ecosystem. In this way, the environmental and ecological relations are studied through reference to conceptually manageable and isolated material parts. After the effective environmental components are understood through reference to their causes; however, they conceptually link back together as an integrated whole, or *holocoenotic system as it was once called. This is known as the* dialectical approach to ecology. The dialectical approach examines the parts, but integrates the organism and the environment into a dynamic whole (or umwelt). Change in one ecological or environmental factor can concurrently affect the dynamic state of an entire ecosystem.

Disturbance and Resilience

Ecosystems are regularly confronted with natural environmental variations and disturbances over time and geographic space. A disturbance is any process that removes biomass from a community, such as a fire, flood, drought, or predation. Disturbances occur over vastly different ranges in terms of magnitudes as well as distances and time periods, and are both the cause and product of natural fluctuations in death rates, species assemblages, and biomass densities within an ecological community. These disturbances create places of renewal where new directions emerge from the patchwork of natural experimentation and opportunity. Ecological resilience is a cornerstone theory in ecosystem management. Biodiversity fuels the resilience of ecosystems acting as a kind of regenerative insurance.

Metabolism and the Early Atmosphere

Metabolism – the rate at which energy and material resources are taken up from the environment, transformed within an organism, and allocated to maintenance, growth and reproduction – is a fundamental physiological trait.

Ernest et al.:991

The Earth was formed approximately 4.5 billion years ago. As it cooled and a crust and oceans formed, its atmosphere transformed from being dominated by hydrogen to one composed mostly of methane and ammonia. Over the next billion years, the metabolic activity of life transformed the atmosphere into a mixture of carbon dioxide, nitrogen, and water vapor. These gases changed the way that light from the sun hit the Earth's surface and greenhouse effects trapped heat. There were untapped sources of free energy within the mixture of reducing and oxidizing gasses that set the stage for primitive ecosystems to evolve and, in turn, the atmosphere also evolved.

The leaf is the primary site of photosynthesis in most plants.

Throughout history, the Earth's atmosphere and biogeochemical cycles have been in a dynamic equilibrium with planetary ecosystems. The history is characterized by periods of significant transformation followed by millions of years of stability. The evolution of the earliest organisms, likely anaerobic methanogen microbes, started the process by converting atmospheric hydrogen into methane ($4H_2 + CO_2 \rightarrow CH_4 + 2H_2O$). Anoxygenic photosynthesis reduced hydrogen concentrations and increased atmospheric methane, by converting hydrogen sulfide into water or other sulfur compounds (for example, $2H_2S + CO_2 + h\upsilon \rightarrow CH_2O + H_2O + 2S$). Early forms of fermentation also increased levels of atmospheric methane. The transition to an oxygen-dominant atmosphere (the *Great Oxidation*) did not begin until approximately 2.4–2.3 billion years ago, but photosynthetic processes started 0.3 to 1 billion years prior.

Radiation: Heat, Temperature and Light

The biology of life operates within a certain range of temperatures. Heat is a form of energy that regulates temperature. Heat affects growth rates, activity, behaviour, and primary production. Temperature is largely dependent on the incidence of solar radiation. The latitudinal and longitudinal spatial variation of temperature greatly affects climates and consequently the distribution of biodiversity and levels of primary production in different ecosystems or biomes across the planet. Heat and temperature relate importantly to metabolic activity. Poikilotherms, for example, have a body temperature that is largely regulated and dependent on the temperature of the external environment. In contrast, homeotherms regulate their internal body temperature by expending metabolic energy.

There is a relationship between light, primary production, and ecological energy budgets. Sunlight is the primary input of energy into the planet's ecosystems. Light is composed of electromagnetic energy of different wavelengths. Radiant energy from the sun generates heat, provides photons of light measured as active energy in the chemical reactions of life, and also acts as a catalyst for genetic mutation. Plants, algae, and some bacteria absorb light and assimilate the energy through photosynthesis. Organisms capable of assimilating energy by photosynthesis or through inorganic fixation of H_2S are autotrophs. Autotrophs — responsible for primary production — assimilate light energy which becomes metabolically stored as potential energy in the form of biochemical enthalpic bonds.

Physical Environments

Water

Wetland conditions such as shallow water, high plant productivity, and anaerobic substrates provide a suitable environment for important physical, biological, and chemical processes. Because of these processes, wetlands play a vital role in global nutrient and element cycles.

Cronk & Fennessy (2001):29

Diffusion of carbon dioxide and oxygen is approximately 10,000 times slower in water than in air. When soils are flooded, they quickly lose oxygen, becoming hypoxic (an environment with O_2 concentration below 2 mg/liter) and eventually completely anoxic where anaerobic bacteria thrive among the roots. Water also influences the intensity and spectral composition of light as it reflects off the water surface and submerged particles. Aquatic plants exhibit a wide variety of morphological and physiological adaptations that allow them to survive, compete, and diversify in these environments. For example, their roots and stems contain large air spaces (aerenchyma) that regulate the efficient transportation of gases (for example, CO_2 and O_2) used in respiration and photosynthesis. Salt water plants (halophytes) have additional specialized adaptations, such as the development of special organs for shedding salt and osmoregulating their internal salt (NaCl) concentrations, to live in estuarine, brackish, or oceanic environments. Anaerobic soil microorganisms in aquatic environments use nitrate, manganese ions, ferric ions, sulfate, carbon dioxide, and some organic compounds; other microorganisms are facultative anaerobes and use oxygen during respiration when the soil becomes drier. The activity of soil microorganisms and the chemistry of the water reduces the oxidation-reduction potentials of the water. Carbon dioxide, for example, is reduced to methane (CH_4) by methanogenic bacteria. The physiology of fish is also specially adapted to compensate for environmental salt levels through osmoregulation. Their gills form electrochemical gradients that mediate salt excretion in salt water and uptake in fresh water.

Gravity

The shape and energy of the land is significantly affected by gravitational forces. On a large scale, the distribution of gravitational forces on the earth is uneven and influences the shape and movement of tectonic plates as well as influencing geomorphic processes such as orogeny and erosion. These forces govern many of the geophysical properties and distributions of ecological biomes across the Earth. On the organismal scale, gravitational forces provide directional cues for plant and fungal growth (gravitropism), orientation cues for animal migrations, and influence the biomechanics and size of animals. Ecological traits, such as allocation of biomass in trees during growth are subject to mechanical failure as gravitational forces influence the position and structure of branches and leaves. The cardiovascular systems of animals are functionally adapted to overcome pressure and gravitational forces that change according to the features of organisms (e.g., height, size, shape), their behaviour (e.g., diving, running, flying), and the habitat occupied (e.g., water, hot deserts, cold tundra).

Pressure

Climatic and osmotic pressure places physiological constraints on organisms, especially those that fly and respire at high altitudes, or dive to deep ocean depths. These constraints influence vertical limits of ecosystems in the biosphere, as organisms are physiologically sensitive and adapted to atmospheric and osmotic water pressure differences. For example, oxygen levels decrease with decreasing pressure and are a limiting factor for life at higher altitudes. Water transportation by plants is another important ecophysiological process affected by osmotic pressure gradients. Water pressure in the depths of oceans requires that organisms adapt to these conditions. For example, diving animals such as whales, dolphins, and seals are specially adapted to deal with changes in sound due to water pressure differences. Differences between hagfish species provide another example of adaptation to deep-sea pressure through specialized protein adaptations.

Turbulent forces in air and water affect the environment and ecosystem distribution, form and dynamics. On a planetary scale, ecosystems are affected by circulation patterns in the global trade winds. Wind power and the turbulent forces it creates can influence heat, nutrient, and biochemical profiles of ecosystems. For example, wind running over the surface of a lake creates turbulence, mixing the water column and influencing the environmental profile to create thermally layered zones, affecting how fish, algae, and other parts of the aquatic ecosystem are structured. Wind speed and turbulence also influence evapotranspiration rates and energy budgets in plants and animals. Wind speed, temperature and moisture content can vary as winds travel across different land features and elevations. For example, the westerlies come into contact with the coastal and interior mountains of western North America to produce a rain shadow on the leeward side of the mountain. The air expands and moisture condenses as the winds increase in elevation; this is called orographic lift and can cause precipitation.[*clarification needed*] This environmental process produces spatial divisions in biodiversity, as species adapted to wetter conditions are range-restricted to the coastal mountain valleys and unable to migrate across the xeric ecosystems (e.g., of the Columbia Basin in western North America) to intermix with sister lineages that are segregated to the interior mountain systems.

Wind and Turbulence

The architecture of the inflorescence in grasses is subject to the physical pressures of wind and shaped by the forces of natural selection facilitating wind-pollination (anemophily).

Fire

Forest fires modify the land by leaving behind an environmental mosaic that diversifies the landscape into different seral stages and habitats of varied quality (left). Some species are adapted to forest fires, such as pine trees that open their cones only after fire exposure (right).

Plants convert carbon dioxide into biomass and emit oxygen into the atmosphere. By approximately 350 million years ago (the end of the Devonian period), photosynthesis had brought the concentration of atmospheric oxygen above 17%, which allowed combustion to occur. Fire releases CO_2 and converts fuel into ash and tar. Fire is a significant ecological parameter that raises many issues pertaining to its control and suppression. While the issue of fire in relation to ecology and plants has been recognized for a long time, Charles Cooper brought attention to the issue of forest fires in relation to the ecology of forest fire suppression and management in the 1960s.

Native North Americans were among the first to influence fire regimes by controlling their spread near their homes or by lighting fires to stimulate the production of herbaceous foods and basketry materials. Fire creates a heterogeneous ecosystem age and canopy structure, and the altered soil nutrient supply and cleared canopy structure opens new ecological niches for seedling establishment. Most ecosystems are adapted to natural fire cycles. Plants, for example, are equipped with a variety of adaptations to deal with forest fires. Some species (e.g., *Pinus halepensis*) *cannot* germinate until after their seeds have lived through a fire or been exposed to certain compounds from smoke. Environmentally triggered germination of seeds is called serotiny. Fire plays a major role in the persistence and resilience of ecosystems.

Soils

Soil is the living top layer of mineral and organic dirt that covers the surface of the planet. It is the chief organizing centre of most ecosystem functions, and it is of critical importance in agricultural science and ecology. The decomposition of dead organic matter (for example, leaves on the forest floor), results in soils containing minerals and nutrients that feed into plant production. The whole of the planet's soil ecosystems is called the pedosphere where a large biomass of the Earth's biodiversity organizes into trophic levels. Invertebrates that feed and shred larger leaves, for example, create smaller bits for smaller organisms in the feeding chain. Collectively, these organisms are the detritivores that regulate soil formation. Tree roots, fungi, bacteria, worms, ants, beetles, centipedes, spiders, mammals, birds, reptiles, amphibians, and other less familiar creatures all work to create the trophic web of life in soil ecosystems. Soils form composite phenotypes where inorganic matter is enveloped into the physiology of a whole community. As organisms feed and migrate through soils they physically displace materials, an ecological process called bioturbation. This aerates soils and stimulates heterotrophic growth and production. Soil microorganisms are influenced by and feed back into the trophic dynamics of the ecosystem. No single axis of causality can be discerned to segregate the biological from geomorphological systems in soils. Paleoecological studies of soils places the origin for bioturbation to a time before the Cambrian period. Other events, such as the evolution of trees and the colonization of land in the Devonian period played a significant role in the early development of ecological trophism in soils.

Biogeochemistry and Climate

Ecologists study and measure nutrient budgets to understand how these materials are regulated, flow, and recycled through the environment. This research has led to an understanding that there is global feedback between ecosystems and the physical parameters of this planet, including minerals, soil, pH, ions, water, and atmospheric gases. Six major elements (hydrogen, carbon, nitrogen, oxygen, sulfur, and phosphorus; H, C, N, O, S, and P) form the constitution of all biological macromolecules and feed into the Earth's geochemical processes. From the smallest scale of biology, the combined effect of billions upon billions of ecological processes amplify and ultimately regulate the biogeochemical cycles of the Earth. Understanding the relations and cycles mediated between these elements and their ecological pathways has significant bearing toward understanding global biogeochemistry.

The ecology of global carbon budgets gives one example of the linkage between biodiversity and biogeochemistry. It is estimated that the Earth's oceans hold 40,000 gigatonnes (Gt) of carbon, that vegetation and soil hold 2070 Gt, and that fossil fuel emissions are 6.3 Gt carbon per year. There have been major restructurings in these global carbon budgets during the Earth's history, regulated to a large extent by the ecology of the land. For example, through the early-mid Eocene volcanic outgassing, the oxidation of methane stored in wetlands, and seafloor gases increased atmospheric CO_2 (carbon dioxide) concentrations to levels as high as 3500 ppm.

In the Oligocene, from twenty-five to thirty-two million years ago, there was another significant restructuring of the global carbon cycle as grasses evolved a new mechanism of photosynthesis, C4 photosynthesis, and expanded their ranges. This new pathway evolved in response to the drop in atmospheric CO_2 concentrations below 550 ppm. The relative abundance and distribution of biodiversity alters the dynamics between organisms and their environment such that ecosystems can be both cause and effect in relation to climate change. Human-driven modifications to the planet's ecosystems (e.g., disturbance, biodiversity loss, agriculture) contributes to rising atmospheric greenhouse gas levels. Transformation of the global carbon cycle in the next century is projected to raise planetary temperatures, lead to more extreme fluctuations in weather, alter species distributions, and increase extinction rates. The effect of global warming is already being registered in melting glaciers, melting mountain ice caps, and rising sea levels. Consequently, species distributions are changing along waterfronts and in continental areas where migration patterns and breeding grounds are tracking the prevailing shifts in climate. Large sections of permafrost are also melting to create a new mosaic of flooded areas having increased rates of soil decomposition activity that raises methane (CH_4) emissions. There is concern over increases in atmospheric methane in the context of the global carbon cycle, because methane is a greenhouse gas that is 23 times more effective at absorbing long-wave radiation than CO_2 on a 100-year time scale. Hence, there is a relationship between global warming, decomposition and respiration in soils and wetlands producing significant climate feedbacks and globally altered biogeochemical cycles.

History

Early Beginnings

Ecology has a complex origin, due in large part to its interdisciplinary nature. Ancient Greek philosophers such as Hippocrates and Aristotle were among the first to record observations on natural history. However, they viewed life in terms of essentialism, where species were conceptualized as static unchanging things while varieties were seen as aberrations of an idealized type. This contrasts against the modern understanding of ecological theory where varieties are viewed as the real phenomena of interest and having a role in the origins of adaptations by means of natural selection. Early conceptions of ecology, such as a balance and regulation in nature can be traced to Herodotus (died *c. 425 BC), who described one of the earliest accounts of* mutualism in his observation of "natural dentistry". Basking Nile crocodiles, he noted, would open their mouths to give sandpipers safe access to pluck leeches out, giving nutrition to the sandpiper and oral hygiene for the crocodile. Aristotle was an early influence on the philosophical development of ecology. He and his student Theophrastus made extensive observations on plant and animal migrations, biogeography, physiology, and on their behaviour, giving an early analogue to the modern concept of an ecological niche.

Ernst Haeckel (left) and Eugenius Warming (right), two founders of ecology

Ecological concepts such as food chains, population regulation, and productivity were first developed in the 1700s, through the published works of microscopist Antoni van Leeuwenhoek (1632–1723) and botanist Richard Bradley (1688?–1732). Biogeographer Alexander von Humboldt (1769–1859) was an early pioneer in ecological thinking and was among the first to recognize ecological gradients, where species are replaced or altered in form along environmental gradients, such as a cline forming along a rise in elevation. Humboldt drew inspiration from Isaac Newton as he developed a form of "terrestrial physics". In Newtonian fashion, he brought a scientific exactitude for measurement into natural history and even alluded to concepts that are the foundation of a modern ecological law on species-to-area relationships. Natural historians, such as Humboldt, James Hutton, and Jean-Baptiste Lamarck (among others) laid the foundations of the modern ecological sciences. The term "ecology" (German: *Oekologie, Ökologie) is of a more recent origin and was first coined by the German biologist* Ernst Haeckel in his book *Generelle Morphologie der Organismen* (1866). Haeckel was a zoologist, artist, writer, and later in life a professor of comparative anatomy.

By ecology, we mean the whole science of the relations of the organism to the environment including, in the broad sense, all the "conditions of existence. "Thus the theory of evolution explains the housekeeping relations of organisms mechanistically as the necessary consequences of effectual causes and so forms the monistic groundwork of ecology.

Ernst Haeckel (1866):140 [B]

Opinions differ on who was the founder of modern ecological theory. Some mark Haeckel's definition as the beginning; others say it was Eugenius Warming with the writing of Oecology of Plants: An Introduction to the Study of Plant Communities (1895), or Carl Linnaeus' principles on the economy of nature that matured in the early 18th century. Linnaeus founded an early branch of ecology that he called the economy of nature. His works influenced Charles Darwin, who adopted Linnaeus' phrase on the *economy or polity of nature in The Origin of Species*. Linnaeus was the first to frame the balance of nature as a testable hypothesis. Haeckel, who admired Darwin's work, defined ecology in reference to the economy of nature, which has led some to question whether ecology and the economy of nature are synonymous.

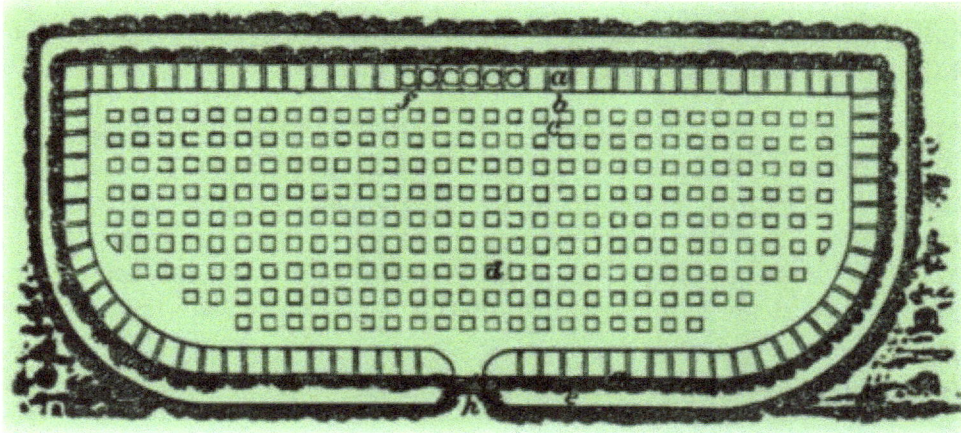

The layout of the first ecological experiment, carried out in a grass garden at Woburn Abbey in 1816, was noted by Charles Darwin in *The Origin of Species*. *The experiment studied the performance of different mixtures of species planted in different kinds of soils.*

From Aristotle until Darwin, the natural world was predominantly considered static and unchanging. Prior to *The Origin of Species, there was little appreciation or understanding of the dynamic and reciprocal relations between organisms, their adaptations, and the environment.* An exception is the 1789 publication *Natural History of Selborne by* Gilbert White (1720–1793), considered by some to be one of the earliest texts on ecology. While Charles Darwin is mainly noted for his treatise on evolution, he was one of the founders of soil ecology, and he made note of the first ecological experiment in *The Origin of Species.* Evolutionary theory changed the way that researchers approached the ecological sciences.

Nowhere can one see more clearly illustrated what may be called the sensibility of such an organic complex,--expressed by the fact that whatever affects any species belonging to it, must speedily have its influence of some sort upon the whole assemblage. He will thus be made to see the impossibility of studying any form completely, out of relation to the other forms,--the necessity for taking a comprehensive survey of the whole as a condition to a satisfactory understanding of any part.

Stephen Forbes (1887)

Since 1900

Modern ecology is a young science that first attracted substantial scientific attention toward the end of the 19th century (around the same time that evolutionary studies were gaining scientific interest). Notable scientist Ellen Swallow Richards may have first introduced the term "oekology" (which eventually morphed into home economics) in the U.S. as early 1892.

In the early 20th century, ecology transitioned from a more descriptive form of natural history to a more analytical form of *scientific natural history*. Frederic Clements published the first American ecology book in 1905, presenting the idea of plant communities as a superorganism. This publication launched a debate between ecological holism and individualism that lasted until the

1970s. Clements' superorganism concept proposed that ecosystems progress through regular and determined stages of seral development that are analogous to the developmental stages of an organism. The Clementsian paradigm was challenged by Henry Gleason, who stated that ecological communities develop from the unique and coincidental association of individual organisms. This perceptual shift placed the focus back onto the life histories of individual organisms and how this relates to the development of community associations.

The Clementsian superorganism theory was an overextended application of an idealistic form of holism. The term "holism" was coined in 1926 by Jan Christiaan Smuts, a South African general and polarizing historical figure who was inspired by Clements' superorganism concept.[C] Around the same time, Charles Elton pioneered the concept of food chains in his classical book *Animal Ecology*. Elton defined ecological relations using concepts of food chains, food cycles, and food size, and described numerical relations among different functional groups and their relative abundance. Elton's 'food cycle' was replaced by 'food web' in a subsequent ecological text. Alfred J. Lotka brought in many theoretical concepts applying thermodynamic principles to ecology.

In 1942, Raymond Lindeman wrote a landmark paper on the trophic dynamics of ecology, which was published posthumously after initially being rejected for its theoretical emphasis. Trophic dynamics became the foundation for much of the work to follow on energy and material flow through ecosystems. Robert MacArthur advanced mathematical theory, predictions, and tests in ecology in the 1950s, which inspired a resurgent school of theoretical mathematical ecologists. Ecology also has developed through contributions from other nations, including Russia's Vladimir Vernadsky and his founding of the biosphere concept in the 1920s and Japan's Kinji Imanishi and his concepts of harmony in nature and habitat segregation in the 1950s. Scientific recognition of contributions to ecology from non-English-speaking cultures is hampered by language and translation barriers.

This whole chain of poisoning, then, seems to rest on a base of minute plants which must have been the original concentrators. But what of the opposite end of the food chain—the human being who, in probable ignorance of all this sequence of events, has rigged his fishing tackle, caught a string of fish from the waters of Clear Lake, and taken them home to fry for his supper?

Rachel Carson (1962):48

Ecology surged in popular and scientific interest during the 1960–1970s environmental movement. There are strong historical and scientific ties between ecology, environmental management, and protection. The historical emphasis and poetic naturalistic writings advocating the protection of wild places by notable ecologists in the history of conservation biology, such as Aldo Leopold and Arthur Tansley, have been seen as far removed from urban centres where, it is claimed, the concentration of pollution and environmental degradation is located. Palamar (2008) notes an overshadowing by mainstream environmentalism of pioneering women in the early 1900s who fought for urban health ecology (then called euthenics) and brought about changes in environmental legislation. Women such as Ellen Swallow Richards and Julia Lathrop, among others, were precursors to the more popularized environmental movements after the 1950s.

In 1962, marine biologist and ecologist Rachel Carson's book *Silent Spring* helped to mobilize the environmental movement by alerting the public to toxic pesticides, such as DDT, bioaccumulating in the environment. Carson used ecological science to link the release of environmental toxins to human and ecosystem health. Since then, ecologists have worked to bridge their understanding of the degradation of the planet's ecosystems with environmental politics, law, restoration, and natural resources management.

Glaciology

Lateral moraine on a glacier joining the Gorner Glacier, Zermatt, Swiss Alps. The moraine is the high bank of debris in the top left hand quarter of the picture. For more explanation, click on the picture.

Glaciology is the scientific study of glaciers, or more generally ice and natural phenomena that involve ice.

Glaciology is an interdisciplinary earth science that integrates geophysics, geology, physical geography, geomorphology, climatology, meteorology, hydrology, biology, and ecology. The impact of glaciers on people includes the fields of human geography and anthropology. The discoveries of water ice on the Moon, Mars, Europa and Pluto add an extraterrestrial component to the field, as in "astroglaciology".

Overview

Areas of study within glaciology include glacial history and the reconstruction of past glaciation. A glaciologist is a person who studies glaciers. Glaciology is one of the key areas of polar research. A glacier is an extended mass of ice formed from snow falling and accumulating over a long period of time; they move very slowly, either descending from high mountains, as in valley glaciers, or moving outward from centers of accumulation, as in continental glaciers.

Types

There are two general categories of glaciation which glaciologists distinguish: *alpine glaciation, accumulations or "rivers of ice"* confined to valleys; and *continental glaciation,* unrestricted accumulations which once covered much of the northern continents.

- Alpine - ice flows down the valleys of mountainous areas and forms a tongue of ice moving towards the plains below. Alpine glaciers tend to make the topography more rugged, by adding and improving the scale of existing features such as large ravines called *cirques and ridges where the rims of two cirques meet called* arêtes.

- Continental - an ice sheet found today, only in high latitudes (Greenland/Antarctica), thousands of square kilometers in area and thousands of meters thick. These tend to smooth out the landscapes.

Glacially-carved Yosemite Valley, as seen from a plane

Zones of Glaciers

- Accumulation, where the formation of ice is faster than its removal.

- Wastage or Ablation, where the sum of melting and evaporation (sublimation) is greater than the amount of snow added each year.

Movement

Ablation

wastage of the glacier through sublimation, ice melting and iceberg calving.

Ablation zone

Area of a glacier in which the annual loss of ice through ablation exceeds the annual gain from precipitation.

Arête

an acute ridge of rock where two cirques abut.

Bergschrund

crevasse formed near the head of a glacier, where the mass of ice has rotated, sheared and torn itself apart in the manner of a geological fault.

Cirque, corrie or cwm

bowl shaped depression excavated by the source of a glacier.

Creep

adjustment to stress at a molecular level.

Flow

movement (of ice) in a constant direction.

Fracture

brittle failure (breaking of ice) under the stress raised when movement is too rapid to be accommodated by creep. It happens for example, as the central part of a glacier moves faster than the edges.

Horn

spire of rock, also known as a pyramidal peak, formed by the headward erosion of three or more cirques around a single mountain. It is an extreme case of an arête.

Plucking/Quarrying

where the adhesion of the ice to the rock is stronger than the cohesion of the rock, part of the rock leaves with the flowing ice.

Tarn

a post-glacial lake in a cirque.

Tunnel valley

The tunnel that is formed by hydraulic erosion of ice and rock below an ice sheet margin. The tunnel valley is what remains of it in the underlying rock when the ice sheet has melted.

Rate of Movement

Moment of the glacier is very slow. It's velocity varies from a few centimeters per day to a few meters per day. The rate of movement depends upon the numbers of factors which are listed below :

- Temperature of the ice
- Gradient of the slope
- Thickness of the glacier

Glacial Deposits

Strati ied Outwash

sand/gravel

> from front of glaciers, found on a plain

Kettles

> block of stagnant ice leaves a depression or pit

Eskers

> steep sided ridges of gravel/sand, possibly caused by streams running under stagnant ice

Kames

> stratified drift builds up low steep hills

Varves

> alternating thin sedimentary beds (coarse and fine) of a proglacial lake. Summer conditions deposit more and coarser material and those of the winter, less and finer.

Unstrati ied

Till-unsorted

> (glacial flour to boulders) deposited by receding/advancing glaciers, forming moraines, and drumlins

Moraines

> (Terminal) material deposited at the end; (Ground) material deposited as glacier melts; (lateral) material deposited along the sides.

Drumlins

> smooth elongated hills composed of till.

Ribbed moraines

> large subglacial elongated hills transverse to former ice flow.

Hydrology

Hydrology is the scientific study of the movement, distribution, and quality of water on Earth and other planets, including the hydrologic cycle, water resources and environmental watershed sustainability. A practitioner of hydrology is a hydrologist, working within the fields of earth or environmental science, physical geography, geology or civil and environmental engineering.

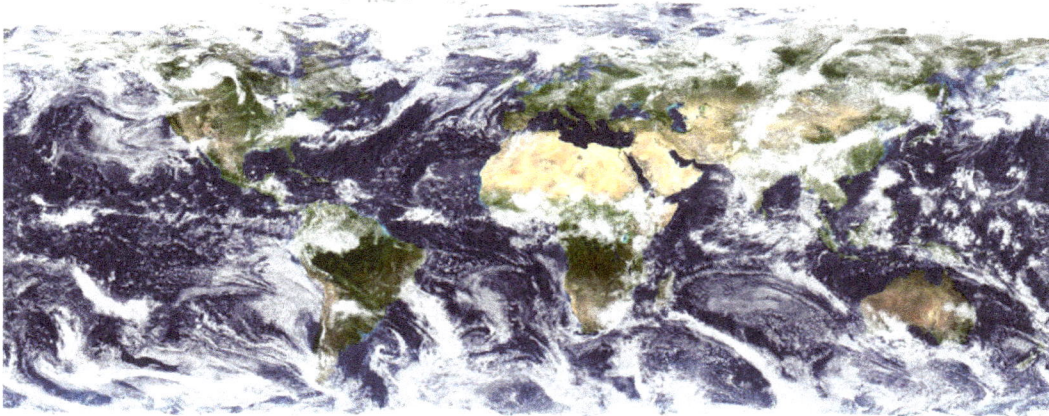

Water covers 70% of the Earth's surface.

Hydrology subdivides into surface water hydrology, groundwater hydrology (hydrogeology), and marine hydrology. Domains of hydrology include hydrometeorology, surface hydrology, hydrogeology, drainage-basin management and water quality, where water plays the central role.

Oceanography and meteorology are not included because water is only one of many important aspects within those fields.

Hydrological research can inform environmental engineering, policy and planning.

History

Hydrology has been a subject of investigation and engineering for millennia. For example, about 4000 BC the Nile was dammed to improve agricultural productivity of previously barren lands. Mesopotamian towns were protected from flooding with high earthen walls. Aqueducts were built by the Greeks and Ancient Romans, while the history of China shows they built irrigation and flood control works. The ancient Sinhalese used hydrology to build complex irrigation works in Sri Lanka, also known for invention of the Valve Pit which allowed construction of large reservoirs, anicuts and canals which still function.

Marcus Vitruvius, in the first century BC, described a philosophical theory of the hydrologic cycle, in which precipitation falling in the mountains infiltrated the Earth's surface and led to streams and springs in the lowlands. With adoption of a more scientific approach, Leonardo da Vinci and Bernard Palissy independently reached an accurate representation of the hydrologic cycle. It was not until the 17th century that hydrologic variables began to be quantified.

Pioneers of the modern science of hydrology include Pierre Perrault, Edme Mariotte and Edmund Halley. By measuring rainfall, runoff, and drainage area, Perrault showed that rainfall was sufficient to account for flow of the Seine. Marriotte combined velocity and river cross-section measurements to obtain discharge, again in the Seine. Halley showed that the evaporation from the Mediterranean Sea was sufficient to account for the outflow of rivers flowing into the sea.

Advances in the 18th century included the Bernoulli piezometer and Bernoulli's equation, by Daniel Bernoulli, and the Pitot tube, by Henri Pitot. The 19th century saw development in groundwater hydrology, including Darcy's law, the Dupuit-Thiem well formula, and Hagen-Poiseuille's capillary flow equation.

Rational analyses began to replace empiricism in the 20th century, while governmental agencies began their own hydrological research programs. Of particular importance were Leroy Sherman's unit hydrograph, the infiltration theory of Robert E. Horton, and C.V. Theis's aquifer test/equation describing well hydraulics.

Since the 1950s, hydrology has been approached with a more theoretical basis than in the past, facilitated by advances in the physical understanding of hydrological processes and by the advent of computers and especially geographic information systems (GIS).

Branches

- Chemical hydrology is the study of the chemical characteristics of water.

- Ecohydrology is the study of interactions between organisms and the hydrologic cycle.

- Hydrogeology is the study of the presence and movement of groundwater.

- Hydroinformatics is the adaptation of information technology to hydrology and water resources applications.

- Hydrometeorology is the study of the transfer of water and energy between land and water body surfaces and the lower atmosphere.

- Isotope hydrology is the study of the isotopic signatures of water.

- Surface hydrology is the study of hydrologic processes that operate at or near Earth's surface.

- Drainage basin management covers water-storage, in the form of reservoirs, and flood-protection.

- Water quality includes the chemistry of water in rivers and lakes, both of pollutants and natural solutes.

Applications

- Determining the water balance of a region.

- Determining the agricultural water balance.

- Designing riparian restoration projects.

- Mitigating and predicting flood, landslide and drought risk.

- Real-time flood forecasting and flood warning.

- Designing irrigation schemes and managing agricultural productivity.

- Part of the hazard module in catastrophe modeling.

- Providing drinking water.

- Designing dams for water supply or hydroelectric power generation.

- Designing bridges.

- Designing sewers and urban drainage system.

- Analyzing the impacts of antecedent moisture on sanitary sewer systems.

- Predicting geomorphologic changes, such as erosion or sedimentation.

- Assessing the impacts of natural and anthropogenic environmental change on water resources.

- Assessing contaminant transport risk and establishing environmental policy guidelines.

Themes

The central theme of hydrology is that water circulates throughout the Earth through different pathways and at different rates. The most vivid image of this is in the evaporation of water from the ocean, which forms clouds. These clouds drift over the land and produce rain. The rainwater flows into lakes, rivers, or aquifers. The water in lakes, rivers, and aquifers then either evaporates back to the atmosphere or eventually flows back to the ocean, completing a cycle. Water changes its state of being several times throughout this cycle.

The areas of research within hydrology concern the movement of water between its various states, or within a given state, or simply quantifying the amounts in these states in a given region. Parts of hydrology concern developing methods for directly measuring these flows or amounts of water, while others concern modelling these processes either for scientific knowledge or for making prediction in practical applications.

Groundwater

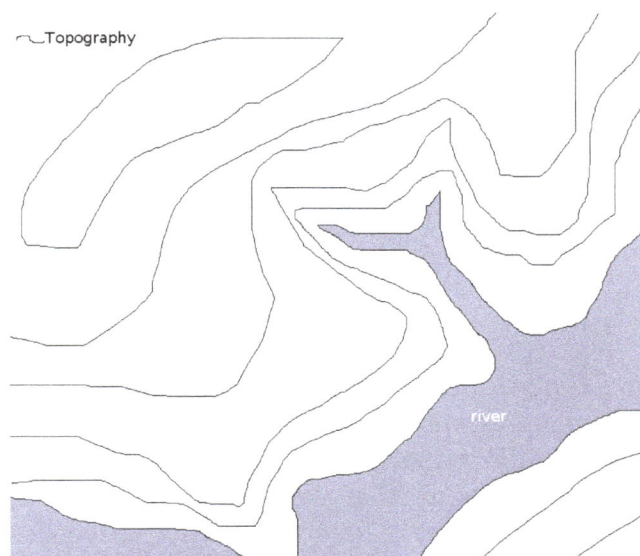

Building a map of groundwater contours

Ground water is water beneath Earth's surface, often pumped for drinking water. Groundwater hydrology (hydrogeology) considers quantifying groundwater flow and solute transport. Problems in describing the saturated zone include the characterization of aquifers in terms of flow direction, groundwater pressure and, by inference, groundwater depth. Measurements here can be made using a piezometer. Aquifers are also described in terms of hydraulic conductivity, storativity and transmisivity. There are a number of geophysical methods for characterising aquifers. There are also problems in characterising the vadose zone (unsaturated zone).

Infiltration

Infiltration is the process by which water enters the soil. Some of the water is absorbed, and the rest percolates down to the water table. The infiltration capacity, the maximum rate at which the soil can absorb water, depends on several factors. The layer that is already saturated provides a resistance that is proportional to its thickness, while that plus the depth of water above the soil provides the driving force (hydraulic head). Dry soil can allow rapid infiltration by capillary action; this force diminishes as the soil becomes wet. Compaction reduces the porosity and the pore sizes. Surface cover increases capacity by retarding runoff, reducing compaction and other processes. Higher temperatures reduce viscosity, increasing infiltration.

Soil Moisture

Soil moisture can be measured in various ways; by capacitance probe, time domain reflectometer or Tensiometer. Other methods include solute sampling and geophysical methods.

Surface Water Flow

Hydrology considers quantifying surface water flow and solute transport, although the treatment of flows in large rivers is sometimes considered as a distinct topic of hydraulics or hydrodynamics. Surface water flow can include flow both in recognizable river channels and otherwise. Methods for measuring flow once water has reached a river include the stream gauge and tracer techniques. Other topics include chemical transport as part of surface water, sediment transport and erosion.

One of the important areas of hydrology is the interchange between rivers and aquifers. Groundwater/surface water interactions in streams and aquifers can be complex and the direction of net water flux (into surface water or into the aquifer) may vary spatially along a stream channel and over time at any particular location, depending on the relationship between stream stage and groundwater levels.

Precipitation and Evaporation

In some considerations, hydrology is thought of as starting at the land-atmosphere boundary and so it is important to have adequate knowledge of both precipitation and evaporation. Precipitation can be measured in various ways: disdrometer for precipitation characteristics at a fine time scale; radar for cloud properties, rain rate estimation, hail and snow detection; Rain gauge for routine accurate measurements of rain and snowfall; satellite – rainy area identification, rain rate estimation, land-cover/land-use, soil moisture.

Evaporation is an important part of the water cycle. It is partly affected by humidity, which can be measured by a sling psychrometer. It is also affected by the presence of snow, hail and ice and can relate to dew, mist and fog. Hydrology considers evaporation of various forms: from water surfaces; as transpiration from plant surfaces in natural and agronomic ecosystems. A direct measurement of evaporation can be obtained using Symon's evaporation pan.

Detailed studies of evaporation involve boundary layer considerations as well as momentum, heat flux and energy budgets.

Remote Sensing

Remote sensing of hydrologic processes can provide information of various types. Sources include land based sensors, airborne sensors and satellite sensors. Information can include clouds, surface moisture, vegetation cover.

Water Quality

In hydrology, studies of water quality concern organic and inorganic compounds, and both dissolved and sediment material. In addition, water quality is affected by the interaction of dissolved oxygen with organic material and various chemical transformations that may take place. Measurements of water quality may involve either in-situ methods, in which analyses take place on-site, often automatically, and laboratory-based analyses and may include microbiological analysis.

Integrating Measurement and Modelling

- Budget analyses
- Parameter estimation
- Scaling in time and space
- Data assimilation
- Quality control of data – see for example Double mass analysis

Prediction

Observations of hydrologic processes are used to make predictions of the future behaviour of hydrologic systems (water flow, water quality). One of the major current concerns in hydrologic research is "Prediction in Ungauged Basins" (PUB), i.e. in basins where no or only very few data exist.

Statistical Hydrology

By analyzing the statistical properties of hydrologic records, such as rainfall or river flow, hydrologists can estimate future hydrologic phenomena. When making assessments of how often relatively rare events will occur, analyses are made in terms of the return period of such events. Other quantities of interest include the average flow in a river, in a year or by season.

These estimates are important for engineers and economists so that proper risk analysis can be performed to influence investment decisions in future infrastructure and to determine the yield reliability characteristics of water supply systems. Statistical information is utilized to formulate operating rules for large dams forming part of systems which include agricultural, industrial and residential demands.

Modeling

Hydrological models are simplified, conceptual representations of a part of the hydrologic cycle. They are primarily used for hydrological prediction and for understanding hydrological processes, within the general field of scientific modeling. Two major types of hydrological models can be distinguished:

- Models based on data. These models are black box systems, using mathematical and statistical concepts to link a certain input (for instance rainfall) to the model output (for instance runoff). Commonly used techniques are regression, transfer functions, and system identification. The simplest of these models may be linear models, but it is common to deploy non-linear components to represent some general aspects of a catchment's response without going deeply into the real physical processes involved. An example of such an aspect is the well-known behavior that a catchment will respond much more quickly and strongly when it is already wet than when it is dry..

- Models based on process descriptions. These models try to represent the physical processes observed in the real world. Typically, such models contain representations of surface runoff, subsurface flow, evapotranspiration, and channel flow, but they can be far more complicated. These models are known as deterministic hydrology models. Deterministic hydrology models can be subdivided into single-event models and continuous simulation models.

Recent research in hydrological modeling tries to have a more global approach to the understanding of the behavior of hydrologic systems to make better predictions and to face the major challenges in water resources management.

Transport

Water movement is a significant means by which other material, such as soil, gravel, boulders or pollutants, are transported from place to place. Initial input to receiving waters may arise from a point source discharge or a line source or area source, such as surface runoff. Since the 1960s rather complex mathematical models have been developed, facilitated by the availability of high speed computers. The most common pollutant classes analyzed are nutrients, pesticides, total dissolved solids and sediment.

Organizations

Intergovernmental Organizations

- International Hydrological Programme (IHP)

International Research Bodies

- International Water Management Institute (IWMI)

- UNESCO-IHE Institute for Water Education

National Research Bodies

- Centre for Ecology and Hydrology – UK

- Centre for Water Science, Cranfield University, UK

- eawag – aquatic research, ETH Zürich, Switzerland

- Institute of Hydrology, Albert-Ludwigs-University of Freiburg, Germany

- United States Geological Survey – Water Resources of the United States

- NOAA's National Weather Service – Office of Hydrologic Development, USA

- US Army Corps of Engineers Hydrologic Engineering Center, USA

- Hydrologic Research Center, USA

- NOAA Economics and Social Sciences, USA

- University of Oklahoma Center for Natural Hazards and Disasters Research, USA

- National Hydrology Research Centre, Canada

- National Institute of Hydrology, India

National and International Societies

- Geological Society of America (GSA) – Hydrogeology Division

- American Geophysical Union (AGU) – Hydrology Section

- National Ground Water Association (NGWA)

- American Water Resources Association

- Consortium of Universities for the Advancement of Hydrologic Science, Inc. (CUAHSI)

- International Association of Hydrological Sciences (IAHS)

- Statistics in Hydrology Working Group (subgroup of IAHS)

- German Hydrological Society (DHG: Deutsche Hydrologische Gesellschaft)

- Italian Hydrological Society (SII-IHS)

- Nordic Association for Hydrology

- British Hydrological Society

- Russian Geographical Society (Moscow Center) – Hydrology Commission

- International Association for Environmental Hydrology

- International Association of Hydrogeologists

Basin- and Catchment-Wide Overviews

- Connected Waters Initiative, University of New South Wales – Investigating and raising awareness of groundwater and water resource issues in Australia

- Murray Darling Basin Initiative, Department of Environment and Heritage, Australia

Limnology

Limnology is the study of inland waters. It is often regarded as a division of ecology or environmental science. It covers the biological, chemical, physical, geological, and other attributes of all inland waters (running and standing waters, both fresh and saline, natural or man-made). This includes the study of lakes and ponds, rivers, springs, streams and wetlands. A more recent sub-disci-pline of limnology, termed landscape limnology, studies, manages, and conserves these aquatic ecosystems using a landscape perspective.

Limnology is closely related to aquatic ecology and hydrobiology, which study aquatic organisms in particular regard to their hydrological environment. Although limnology is sometimes equated with freshwater science, this is erroneous since limnology also comprises the study of inland salt lakes.

Lake Hāwea, New Zealand

History

The term limnology was coined by François-Alphonse Forel (1841–1912) who established the field with his studies of Lake Geneva. Interest in the discipline rapidly expanded, and in 1922 August Thienemann (a German zoologist) and Einar Naumann (a Swedish botanist) co-founded the International Society of Limnology (SIL, from Societas Internationalis Limnologiae). Forel's original definition of limnology, "the oceanography of lakes", was expanded to encompass the study of all inland waters, and influenced Benedykt Dybowski's work on Lake Baikal.

Prominent early American limnologists included G. Evelyn Hutchinson, Ed Deevey, E. A. Birge, and C. Juday.

Lake Classification

Lake George, New York, United States, an oligotrophic lake

Limnology classifies lakes (or other bodies of water) according to the trophic state index. An oligotrophic lake is characterised by relatively low levels of primary production and low levels of nutrients. A eutrophic lake has high levels of primary productivity due to very high nutrient levels. Eutrophication of a lake can lead to algal blooms. Dystrophic lakes have high levels of humic matter and typically has yellow-brown, tea-coloured waters. These categories do not have rigid specifications; the classification system can be seen as more of a spectrum encompassing the various levels of aquatic productivity.

Organizations

- Association for the Sciences of Limnology and Oceanography
- Asociación Ibérica de Limnología
- Australian Society for Limnology
- Society of Canadian Limnologists
- European Society of Limnology and Oceanography
- Society of Limnology
- Italian Association for Oceanology and Limnology (AIOL)
- The Japanese Society of Limnology
- International Society of Limnology
- Brazilian Society of Limnology
- New Zealand freshwater Sciences society
- Southern African Society of Aquatic Scientists
- Balaton Limnological Institute
- Polish Limnological Society
- Society for Freshwater Science (formerly North American Benthological Society)
- Israel Oceanographic and Limnological Research
- Czech Limnological Society
- Freshwater Biological Association (UK)

Seismology

Seismology is the scientific study of earthquakes and the propagation of elastic waves through the Earth or through other planet-like bodies. The field also includes studies of earthquake environmental effects, such as tsunamis as well as diverse seismic sources such as volcanic, tectonic, oceanic, atmospheric, and artificial processes (such as explosions). A related field that uses geology to infer information regarding past earthquakes is paleoseismology. A re-cording of earth motion as a function of time is called a seismogram. A seismologist is a scientist who does research in seismology.

History

Scholarly interest in earthquakes can be traced back to antiquity. Early speculations on the natural causes of earthquakes were included in the writings of Thales of Miletus (c. 585 BCE), Anaximenes of Miletus (c. 550 BCE), Aristotle (c. 340 BCE) and Zhang Heng (132 CE).

In 132 CE, Zhang Heng of China's Han dynasty designed the first known seismoscope.

In 1664, Athanasius Kircher argued that earthquakes were caused by the movement of fire within a system of channels inside the Earth.

In 1703, Martin Lister (1638 to 1712) and Nicolas Lemery (1645 to 1715) proposed that earthquakes were caused by chemical explosions within the earth.

The Lisbon earthquake of 1755, coinciding with the general flowering of science in Europe, set in motion intensified scientific attempts to understand the behaviour and causation of earthquakes. The earliest responses include work by John Bevis (1757) and John Michell (1761). Michell determined that earthquakes originate within the Earth and were waves of movement caused by "shifting masses of rock miles below the surface."

From 1857, Robert Mallet laid the foundation of instrumental seismology and carried out seismological experiments using explosives. He is also responsible for coining the word "seismology".

In 1897, Emil Wiechert's theoretical calculations led him to conclude that the Earth's interior consists of a mantle of silicates, surrounding a core of iron.

In 1906 Richard Dixon Oldham identified the separate arrival of P-waves, S-waves and surface waves on seismograms and found the first clear evidence that the Earth has a central core.

In 1910, after studying the 1906 San Francisco earthquake, Harry Fielding Reid put forward the "elastic rebound theory" which remains the foundation for modern tectonic studies. The development of this theory depended on the considerable progress of earlier independent streams of work on the behaviour of elastic materials and in mathematics.

In 1926, Harold Jeffreys was the first to claim, based on his study of earthquake waves, that below the crust, the core of the Earth is liquid.

In 1937, Inge Lehmann determined that within the earth's liquid outer core there is a solid *inner core*.

By the 1960s, earth science had developed to the point where a comprehensive theory of the causation of seismic events had come together in the now well-established theory of plate tectonics.

Types of Seismic Wave

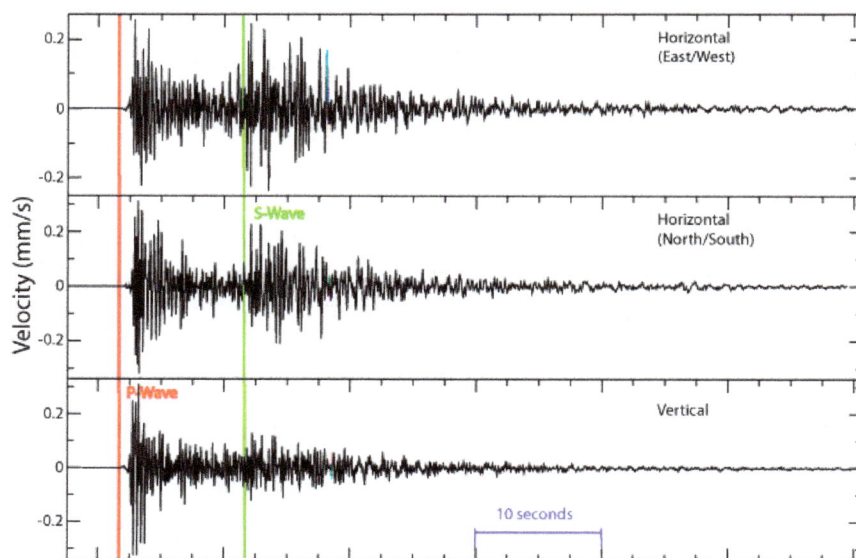

Seismogram records showing the three components of ground motion. The red line marks the first arrival of P-waves; the green line, the later arrival of S-waves.

Seismic waves are elastic waves that propagate in solid or fluid materials. They can be divided into body waves that travel through the interior of the materials; surface waves that travel along surfaces or interfaces between materials; and normal modes, a form of standing wave.

Body Waves

There are two types of body waves, Pressure waves or Primary waves (P-waves) and Shear or Secondary waves (S-waves). P-waves, are longitudinal waves that involve compression and expansion in the direction that the wave is moving. P-waves are the fastest waves in solids and are therefore the first waves to appear on a seismogram. S-waves are transverse waves that move perpendicular to the direction of propagation. S-waves are slower than P-waves. Therefore, they appear later than P-waves on a seismogram. Fluids cannot support perpendicular motion, so S-waves only travel in solids.

Surface Waves

The two main surface wave types are Rayleigh waves, which have some compressional motion, and Love waves, which do not. Rayleigh waves result from the interaction of vertically polarized P- and S-waves that satisfy the boundary conditions on the surface. Love waves can exist in the presence of a subsurface layer, and are only formed by horizontally polarized S-waves. Surface waves travel more slowly than P-waves and S-waves; however, because they are guided by the Earth's surface and their energy is thus trapped near the surface, they can be much stronger than body waves, and can be the largest signals on earthquake seismograms. Surface waves are strongly excited when their source is close to the surface, as in a shallow earthquake or a near surface explosion.

Normal Modes

Both body and surface waves are traveling waves; however, large earthquakes can also make the Earth "ring" like a bell. This ringing is a mixture of normal modes with discrete frequencies and periods of an hour or shorter. Motion caused by a large earthquake can be observed for up to a month after the event. The first observations of normal modes were made in the 1960s as the advent of higher fidelity instruments coincided with two of the largest earthquakes of the 20th century - the 1960 Valdivia earthquake and the 1964 Alaska earthquake. Since then, the normal modes of the Earth have given us some of the strongest constraints on the deep structure of the Earth.

Earthquakes

One of the first attempts at the scientific study of earthquakes followed the 1755 Lisbon earthquake. Other notable earthquakes that spurred major advancements in the science of seismology include the 1857 Basilicata earthquake, 1906 San Francisco earthquake, the 1964 Alaska earthquake, the 2004 Sumatra-Andaman earthquake, and the 2011 Great East Japan earthquake.

Controlled Seismic Sources

Seismic waves produced by explosions or vibrating controlled sources are one of the primary methods of underground exploration in geophysics (in addition to many different electromagnetic methods such as induced polarization and magnetotellurics). Controlled-source seismology has been used to map salt domes, anticlines and other geologic traps in petroleum-bearing rocks, faults, rock types, and long-buried giant meteor craters. For example, the Chicxulub Crater, which was caused by an impact that has been implicated in the extinction of the dinosaurs, was localized to Central America by analyzing ejecta in the Cretaceous–Paleogene boundary, and then physically proven to exist using seismic maps from oil exploration.

Detection of Seismic Waves

Seismometers are sensors that sense and record the motion of the Earth arising from elastic waves. Seismometers may be deployed at the Earth's surface, in shallow vaults, in boreholes, or underwater. A complete instrument package that records seismic signals is called a seismograph. Networks of seismographs continuously record ground motions around the world to facilitate the monitoring and analysis of global earthquakes and other sources of seismic activity. Rapid location of earthquakes makes tsunami warnings possible because seismic waves travel considerably faster than tsunami waves. Seismometers also record signals from non-earthquake sources ranging from explosions (nuclear and chemical), to local noise from wind or anthropogenic activities, to incessant signals generated at the ocean floor and coasts induced by ocean waves (the global microseism), to cryospheric events associated with large icebergs and glaciers. Above-ocean meteor strikes with energies as high as 4.2×10^{13} J (equivalent to that released by an explosion of ten kilotons of TNT) have been recorded by seismographs, as have a number of industrial accidents and terrorist bombs and events (a field of study referred to as forensic seismology). A major long-term motivation for the global seismographic monitoring has been for the detection and study of nuclear testing.

Installation for a temporary seismic station, north Iceland highland.

Mapping the Earth's Interior

Because seismic waves commonly propagate efficiently as they interact with the internal structure of the Earth, they provide high-resolution noninvasive methods for studying the planet's interior. One of the earliest important discoveries (suggested by Richard Dixon Oldham in 1906 and definitively shown by Harold Jeffreys in 1926) was that the outer core of the earth is liquid. Since S-waves do not pass through liquids, the liquid core causes a "shadow" on the side of the planet opposite of the earthquake where no direct S-waves are observed. In addition, P-waves travel much slower through the outer core than the mantle.

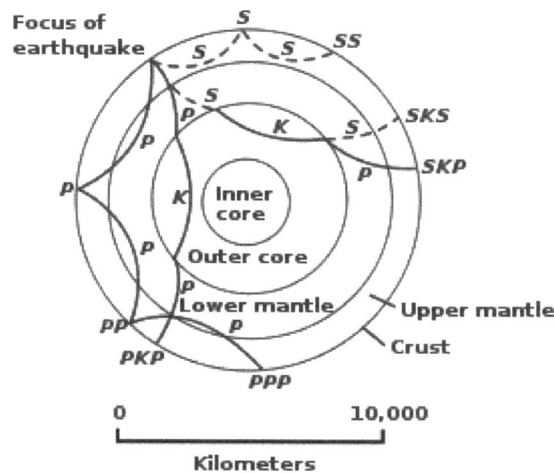

Seismic velocities and boundaries in the interior of the Earth sampled by seismic waves

Processing readings from many seismometers using seismic tomography, seismologists have mapped the mantle of the earth to a resolution of several hundred kilometers. This has enabled scientists to identify convection cells and other large-scale features such as the large low-shear-velocity provinces near the core–mantle boundary.

Seismology and Society

Earthquake Prediction

Forecasting a probable timing, location, magnitude and other important features of a forthcoming seismic event is called earthquake prediction. Various attempts have been made by seismologists and others to create effective systems for precise earthquake predictions, including the VAN method. Most seismologists do not believe that a system to provide timely warnings for individual earthquakes has yet been developed, and many believe that such a system would be unlikely to give useful warning of impending seismic events. However, more general forecasts routinely predict seismic hazard. Such forecasts estimate the probability of an earthquake of a particular size affecting a particular location within a particular time-span, and they are routinely used in earthquake engineering.

Public controversy over earthquake prediction erupted after Italian authorities indicted six seismologists and one government official for manslaughter in connection with a magnitude 6.3 earthquake in L'Aquila, Italy on April 5, 2009. The indictment has been widely perceived[*by whom?*] as an indictment for failing to predict the earthquake and has drawn condemnation from the American Association for the Advancement of Science and the American Geophysical Union. The indictment claims that, at a special meeting in L'Aquila the week before the earthquake occurred, scientists and officials were more interested in pacifying the population than providing adequate information about earthquake risk and preparedness.

Engineering Seismology

Engineering seismology is the study and application of seismology for engineering purposes. It generally applied to the branch of seismology that deals with the assessment of the seismic hazard of a site or region for the purposes of earthquake engineering. It is, therefore, a link between earth science and civil engineering. There are two principal components of engineering seismology. Firstly, studying earthquake history (e.g. historical and instrumental catalogs of seismicity) and tectonics to assess the earthquakes that could occur in a region and their characteristics and frequency of occurrence. Secondly, studying strong ground motions generated by earthquakes to assess the expected shaking from future earthquakes with similar characteristics. These strong ground motions could either be observations from accelerometers or seismometers or those simulated by computers using various techniques.

Tools

Seismological instruments can generate large amounts of data. Systems for processing such data include:

- CUSP (Caltech-USGS Seismic Processing)

- RadExPro seismic software

- SeisComP3

Stratigraphy

The Permian through Jurassic strata of the Colorado Plateau area of southeastern Utah demonstrate the principles of stratigraphy.

Stratigraphy is a branch of geology which studies rock layers (strata) and layering (stratification). It is primarily used in the study of sedimentary and layered volcanic rocks. Stratigraphy includes two related subfields: lithologic stratigraphy or lithostratigraphy, and biologic stratigraphy or biostratigraphy.

Historical Development

Nicholas Steno established the theoretical basis for stratigraphy when he introduced the law of superposition, the principle of original horizontality and the principle of lateral continuity in a 1669 work on the fossilization of organic remains in layers of sediment.

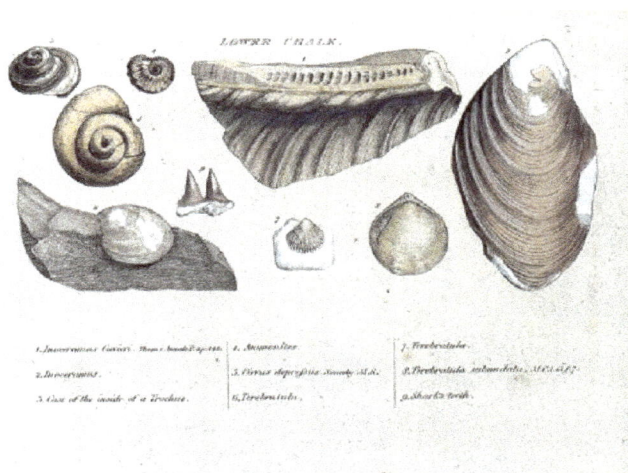

Engraving from William Smith's monograph on identifying strata based on fossils

The first practical large-scale application of stratigraphy was by William Smith in the 1790s and early 19th century. Smith, known as the "Father of English geology", created the first geologic map of England and first recognized the significance of strata or rock layering and the importance of fossil markers for correlating strata. Another influential application of stratigraphy in the early 19th century was a study by Georges Cuvier and Alexandre Brongniart of the geology of the region around Paris.

Strata in Cafayate (Argentina)

Lithostratigraphy

Lithostratigraphy, or lithologic stratigraphy, provides the most obvious visible layering. It deals with the physical contrasts in lithology, or rock type. Such layers can occur both vertically – in layering or bedding of varying rock types – and laterally – reflecting changing environments of deposition (known as facies change). Key concepts in stratigraphy involve understanding how certain geometric relationships between rock layers arise and what these geometries mean in terms of the depositional environment. Stratigraphers have codified a basic concept of their discipline in the law of superposition, which simply states that, in an undeformed stratigraphic sequence, the oldest strata occur at the base of the sequence.

Chalk layers in Cyprus, showing sedimentary layering

Chemostratigraphy studies the changes in the relative proportions of trace elements and isotopes within and between lithologic units. Carbon and oxygen isotope ratios vary with time, and researchers can use them to map subtle changes that occurred in the paleoenvironment. This has led to the specialized field of isotopic stratigraphy.

Cyclostratigraphy documents the often cyclic changes in the relative proportions of minerals (particularly carbonates), grain size, or thickness of sediment layers (varves) and of fossil diversity with time, related to seasonal or longer term changes in palaeoclimates.

Biostratigraphy

Biostratigraphy or paleontologic stratigraphy is based on fossil evidence in the rock layers. Strata from widespread locations containing the same fossil fauna and flora are correlatable in time. Biologic stratigraphy was based on William Smith's principle of faunal succession, which predated, and was one of the first and most powerful lines of evidence for, biological evolution. It provides strong evidence for the formation (speciation) and extinction of species. The geologic time scale was developed during the 19th century, based on the evidence of biologic stratigraphy and faunal succession. This timescale remained a relative scale until the development of radiometric dating, which gave it and the stratigraphy it was based on an absolute time framework, leading to the development of chronostratigraphy.

One important development is the Vail curve, which attempts to define a global historical sea-level curve according to inferences from worldwide stratigraphic patterns. Stratigraphy is also commonly used to delineate the nature and extent of hydrocarbon-bearing reservoir rocks, seals, and traps in petroleum geology.

Chronostratigraphy

Chronostratigraphy is the branch of stratigraphy that studies the absolute, not relative, age of rock strata. The branch is concerned with deriving geochronological data for rock units, both directly and inferentially, so that a sequence of time-relative events of rocks within a region can be derived. In essence, chronostratigraphy seeks to understand the geologic history of rocks and regions.

The ultimate aim of chronostratigraphy is to arrange the sequence of deposition and the time of deposition of all rocks within a geological region and, eventually, the entire geologic record of the earth.

A gap or missing strata in the geological record of an area is called a stratigraphic hiatus. This may be the result of lack of sediment deposition or it may be due to removal by erosion, in which case it may be called a vacuity. It is called a *hiatus because deposition was on hold for a period of time.* A physical gap may represent both a period of non-deposition and a period of erosion. A fault may cause the appearance of a hiatus.

Magnetostratigraphy

Magnetostratigraphy is a chronostratigraphic technique used to date sedimentary and volcanic sequences. The method works by collecting oriented samples at measured intervals throughout a section. The samples are analyzed to determine their detrital remanent magnetism (DRM), that is, the polarity of Earth's magnetic field at the time a stratum was deposited. For sedimentary rocks, this is possible because, when very fine-grained magnetic minerals (< 17 μm) fall through the water column, they orient themselves with Earth's magnetic field. Upon burial, that orientation is preserved. The minerals behave like tiny compasses. For volcanic rocks, magnetic minerals, which form in the melt, are fixed in place upon crystallization or freezing of the lava and are oriented with the ambient magnetic field.

Oriented paleomagnetic core samples are collected in the field; mudstones, siltstones, and very fine-grained sandstones are the preferred lithologies because the magnetic grains are finer and more likely to orient with the ambient field during deposition. If the ancient magnetic field were oriented similar to today's field (North Magnetic Pole near the North Rotational Pole), the strata would retain a normal polarity. If the data indicate that the North Magnetic Pole were near the South Rotational Pole, the strata would exhibit reversed polarity.

Results of the individual samples are analyzed by removing the natural remanent magnetization (NRM) to reveal the DRM. Following statistical analysis, the results are used to generate a local magnetostratigraphic column that can then be compared against the Global Magnetic Polarity Time Scale.

This technique is used to date sequences that generally lack fossils or interbedded igneous rocks. The continuous nature of the sampling means that it is also a powerful technique for the estimation of sediment-accumulation rates.

Archaeological Stratigraphy

In the field of archaeology, soil stratigraphy is used to better understand the processes that form and protect archaeological sites. Since the law of superposition holds true, it can help date finds or features from each context; these finds and features can be placed in sequence and the dates interpolated. Phases of activity can also often be seen through stratigraphy, especially when a trench or feature is viewed in section (profile). Because pits and other features can be dug down into earlier levels, not all material at the same absolute depth is necessarily of the same age; close attention has to be paid to the archeological layers. The Harris-matrix is a tool to depict complex stratigraphic relations when they are found, for example, in the context of urban archaeology.

References

- "What is planetary geology?". James F. Bell III (Cornell University), Bruce A. Campbell (Smithsonian Institution), Mark S. Robinson (U.S. Geological Survey). Retrieved 6 October 2015.

- Williamson, Fiona (2015-09-01). "Weathering the empire: meteorological research in the early British straits settlements". The British Journal for the History of Science. 48 (03): 475–492. doi:10.1017/S000708741500028X. ISSN 1474-001X.

- Glickman, Todd S. (June 2009). Meteorology Glossary (electronic) (2nd ed.). Cambridge, Massachusetts: American Meteorological Society. Retrieved March 10, 2014.

- Glickman, Todd S. (June 2000). Meteorology Glossary (electronic) (2nd ed.). Cambridge, Massachusetts: American Meteorological Society. Retrieved March 10, 2014.

- "Integrated Water Resource Management in Australia: Case studies – Murray-Darling Basin initiative". Australian Government, Department of the Environment. Australian Government. Retrieved 19 June 2014.

- A. J. Bowden; Cynthia V. Burek; C. V. Burek; Richard Wilding (2005). History of palaeobotany: selected essays. Geological Society. p. 293. ISBN 978-1-86239-174-1. Retrieved 3 April 2013.

- "Eawag aquatic research". Swiss Federal Institute of Aquatic Science and Technology. 25 January 2012. Retrieved 8 March 2013.

- "National Hydrology Research Centre (Saskatoon, SK)". Environmental Science Centres. Environment Canada. Retrieved 8 March 2013.

- Al-Rawi, Munin M. (November 2002). The Contribution of Ibn Sina (Avicenna) to the development of Earth Sciences (PDF) (Report). Manchester, UK: Foundation for Science Technology and Civilisation. Publication 4039. Retrieved April 2012.

Applications of Earth and Planetary Science

The Earth is not a monolith and it has gone through many changes in its atmosphere, climate, as well as geological structures. These changes have been faithfully preserved in geologic and other fossilized records that are available to scientists. They help to understand and predict changes in climate, seismic activities and volcanic activities of the Earth. The topics discussed in the chapter are of great importance to broaden the existing knowledge on earth and planetary science.

Geophysics

Geophysics is a subject of natural science concerned with the physical processes and physical properties of the Earth and its surrounding space environment, and the use of quantitative methods for their analysis. The term *geophysics* sometimes refers only to the geological applications: Earth's shape; its gravitational and magnetic fields; its internal structure and composition; its dynamics and their surface expression in plate tectonics, the generation of magmas, volcanism and rock formation. However, modern geophysics organizations use a broader definition that includes the water cycle including snow and ice; fluid dynamics of the oceans and the atmosphere; electricity and magnetism in the ionosphere and magnetosphere and solar-terrestrial relations; and analogous problems associated with the Moon and other planets.

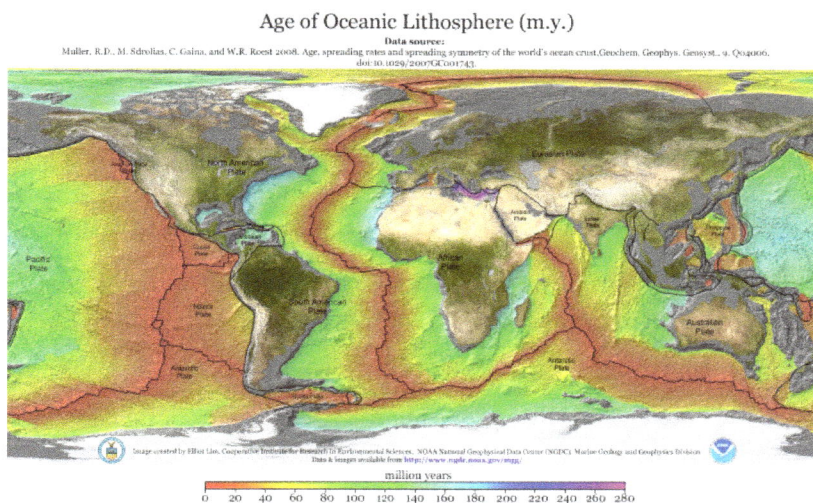

Age of Oceanic Lithosphere (m.y.)

Age of the sea floor. Much of the dating information comes from magnetic anomalies.

Although geophysics was only recognized as a separate discipline in the 19th century, its origins date back to ancient times. The first magnetic compasses were made from lodestones, while more modern magnetic compasses played an important role in the history of navigation. The first seismic instrument was built in 132 BC. Isaac Newton applied his theory of mechanics to the tides and the precession of the equinox; and instruments were developed to measure the Earth's shape, density and gravity field, as well as the components of the water cycle. In the 20th century, geophysical methods were developed for remote exploration of the solid Earth and the ocean, and geophysics played an essential role in the development of the theory of plate tectonics.

Geophysics is applied to societal needs, such as mineral resources, mitigation of natural hazards and environmental protection. Geophysical survey data are used to analyze potential petroleum reservoirs and mineral deposits, locate groundwater, find archaeological relics, determine the thickness of glaciers and soils, and assess sites for environmental remediation.

Physical Phenomena

Geophysics is a highly interdisciplinary subject, and geophysicists contribute to every area of the Earth sciences. To provide a clearer idea of what constitutes geophysics, this section describes phenomena that are studied in physics and how they relate to the Earth and its surroundings.

Gravity

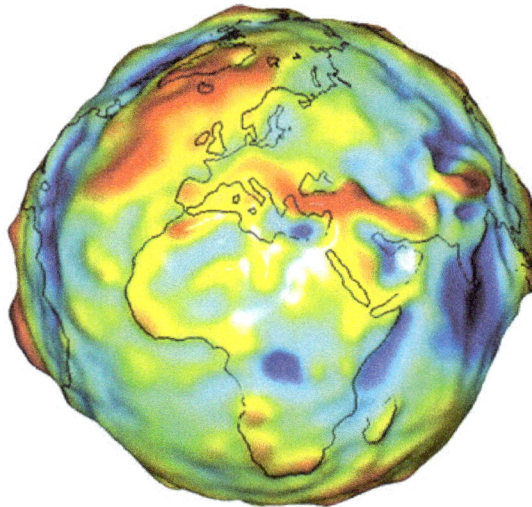

A map of deviations in gravity from a perfectly smooth, idealized Earth.

The gravitational pull of the Moon and Sun give rise to two high tides and two low tides every lunar day, or every 24 hours and 50 minutes. Therefore, there is a gap of 12 hours and 25 minutes between every high tide and between every low tide.

Gravitational forces make rocks press down on deeper rocks, increasing their density as the depth increases. Measurements of gravitational acceleration and gravitational potential at the Earth's surface and above it can be used to look for mineral deposits. The surface gravitational field provides information on the dynamics of tectonic plates. The geopotential surface called the geoid is one definition of the shape of the Earth. The geoid would be the global mean sea level if the oceans were in equilibrium and could be extended through the continents (such as with very narrow canals).

Heat Flow

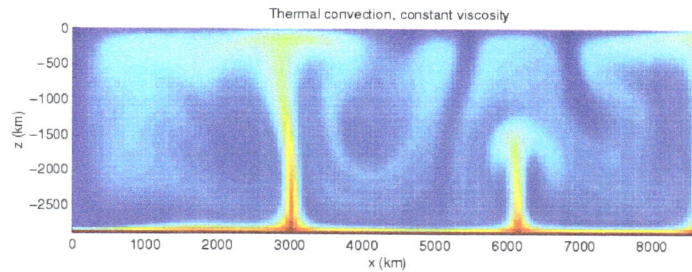

Thermal convection, constant viscosity

A model of thermal convection in the Earth's mantle. The thin red columns are mantle plumes.

The Earth is cooling, and the resulting heat flow generates the Earth's magnetic field through the geodynamo and plate tectonics through mantle convection. The main sources of heat are the primordial heat and radioactivity, although there are also contributions from phase transitions. Heat is mostly carried to the surface by thermal convection, although there are two thermal boundary layers – the core-mantle boundary and the lithosphere – in which heat is transported by conduction. Some heat is carried up from the bottom of the mantle by mantle plumes. The heat flow at the Earth's surface is about 4.2×10^{13} W, and it is a potential source of geothermal energy.

Vibrations

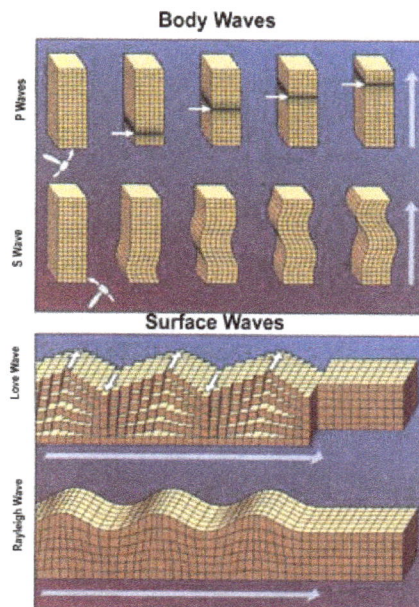

Illustration of the deformations of a block by body waves and surface waves.

Seismic waves are vibrations that travel through the Earth's interior or along its surface. The entire Earth can also oscillate in forms that are called normal modes or free oscillations of the Earth. Ground motions from waves or normal modes are measured using seismographs. If the waves come from a localized source such as an earthquake or explosion, measurements at more than one location can be used to locate the source. The locations of earthquakes provide information on plate tectonics and mantle convection.

Measurements of seismic waves are a source of information on the region that the waves travel through. If the density or composition of the rock changes suddenly, some waves are reflected. Reflections can provide information on near-surface structure. Changes in the travel direction, called refraction, can be used to infer the deep structure of the Earth.

Earthquakes pose a risk to humans. Understanding their mechanisms, which depend on the type of earthquake (e.g., intraplate or deep focus), can lead to better estimates of earthquake risk and improvements in earthquake engineering.

Electricity

Although we mainly notice electricity during thunderstorms, there is always a downward electric field near the surface that averages 120 V m−1. Relative to the solid Earth, the atmosphere has a net positive charge due to bombardment by cosmic rays. A current of about 1800 A flows in the global circuit. It flows downward from the ionosphere over most of the Earth and back upwards through thunderstorms. The flow is manifested by lightning below the clouds and sprites above.

A variety of electric methods are used in geophysical survey. Some measure spontaneous potential, a potential that arises in the ground because of man-made or natural disturbances. Telluric currents flow in Earth and the oceans. They have two causes: electromagnetic induction by the time-varying, external-origin geomagnetic field and motion of conducting bodies (such as seawater) across the Earth's permanent magnetic field. The distribution of telluric current density can be used to detect variations in electrical resistivity of underground structures. Geophysicists can also provide the electric current themselves.

Electromagnetic Waves

Electromagnetic waves occur in the ionosphere and magnetosphere as well as the Earth's outer core. Dawn chorus is believed to be caused by high-energy electrons that get caught in the Van Allen radiation belt. Whistlers are produced by lightning strikes. Hiss may be generated by both. Electromagnetic waves may also be generated by earthquakes.

In the Earth's outer core, electric currents in the highly conductive liquid iron create magnetic fields by electromagnetic induction. Alfvén waves are magnetohydrodynamic waves in the magnetosphere or the Earth's core. In the core, they probably have little observable effect on the geomagnetic field, but slower waves such as magnetic Rossby waves may be one source of geomagnetic secular variation.

Electromagnetic methods that are used for geophysical survey include transient electromagnetics and magnetotellurics·

Magnetism

The Earth's magnetic field protects the Earth from the deadly solar wind and has long been used for navigation. It originates in the fluid motions of the Earth's outer core. The magnetic field in the upper atmosphere gives rise to the auroras.

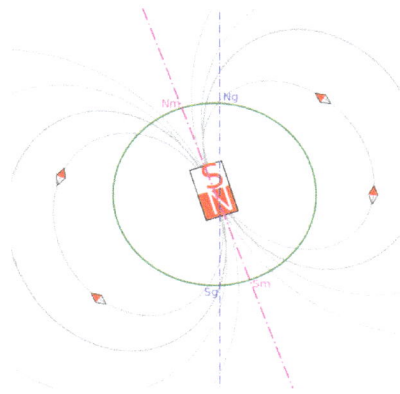

Earth's dipole axis (pink line) is tilted away from the rotational axis (blue line).

The Earth's field is roughly like a tilted dipole, but it changes over time (a phenomenon called geomagnetic secular variation). Mostly the geomagnetic pole stays near the geographic pole, but at random intervals averaging 440,000 to a million years or so, the polarity of the Earth's field reverses. These geomagnetic reversals, analyzed within a Geomagnetic Polarity Time Scale, contain 184 polarity intervals in the last 83 million years, with change in frequency over time, with the most recent brief complete reversal of the Laschamp event occurring 41,000 years ago during the last glacial period. Geologists observed geomagnetic reversal recorded in volcanic rocks, through magnetostratigraphy correlation and their signature can be seen as parallel linear magnetic anomaly stripes on the seafloor. These stripes provide quantitative information on seafloor spreading, a part of plate tectonics. They are the basis of magnetostratig-raphy, which correlates magnetic reversals with other stratigraphies to construct geologic time scales. In addition, the magnetization in rocks can be used to measure the motion of continents.

Radioactivity

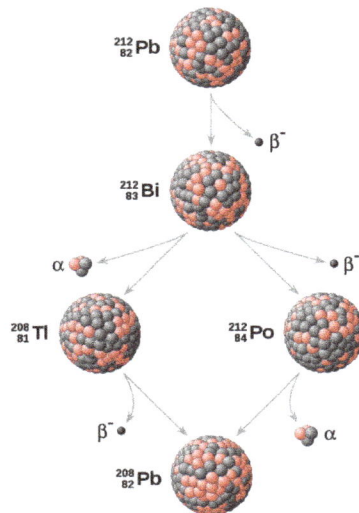

Example of a radioactive decay chain.

Radioactive decay accounts for about 80% of the Earth's internal heat, powering the geodyna-mo and plate tectonics. The main heat-producing isotopes are potassium-40, uranium-238, uranium-235, and thorium-232. Radioactive elements are used for radiometric dating, the primary method for establishing an absolute time scale in geochronology. Unstable isotopes decay at predictable rates, and the decay rates of different isotopes cover several orders of magnitude, so radioactive decay can be used to accurately date both recent events and events in past geologic eras. Radiometric mapping using ground and airborne gamma spectrometry can be used to map the concentration and distribution of radioisotopes near the Earth's surface, which is useful for mapping lithology and alteration.

Fluid Dynamics

Fluid motions occur in the magnetosphere, atmosphere, ocean, mantle and core. Even the mantle, though it has an enormous viscosity, flows like a fluid over long time intervals. This flow is reflected in phenomena such as isostasy, post-glacial rebound and mantle plumes. The mantle flow drives plate tectonics and the flow in the Earth's core drives the geodynamo.

Geophysical fluid dynamics is a primary tool in physical oceanography and meteorology. The rotation of the Earth has profound effects on the Earth's fluid dynamics, often due to the Coriolis effect. In the atmosphere it gives rise to large-scale patterns like Rossby waves and determines the basic circulation patterns of storms. In the ocean they drive large-scale circulation patterns as well as Kelvin waves and Ekman spirals at the ocean surface. In the Earth's core, the circulation of the molten iron is structured by Taylor columns.

Waves and other phenomena in the magnetosphere can be modeled using magnetohydrodynamics.

Mineral Physics

The physical properties of minerals must be understood to infer the composition of the Earth's interior from seismology, the geothermal gradient and other sources of information. Mineral physicists study the elastic properties of minerals; their high-pressure phase diagrams, melting points and equations of state at high pressure; and the rheological properties of rocks, or their ability to flow. Deformation of rocks by creep make flow possible, although over short times the rocks are brittle. The viscosity of rocks is affected by temperature and pressure, and in turn determines the rates at which tectonic plates move.

Water is a very complex substance and its unique properties are essential for life. Its physical properties shape the hydrosphere and are an essential part of the water cycle and climate. Its thermodynamic properties determine evaporation and the thermal gradient in the atmosphere. The many types of precipitation involve a complex mixture of processes such as coalescence, supercooling and supersaturation. Some precipitated water becomes groundwater, and groundwater flow includes phenomena such as percolation, while the conductivity of water makes electrical and electromagnetic methods useful for tracking groundwater flow. Physical properties of water such as salinity have a large effect on its motion in the oceans.

The many phases of ice form the cryosphere and come in forms like ice sheets, glaciers, sea ice, freshwater ice, snow, and frozen ground (or permafrost).

Regions of the Earth

Size and form of the Earth

The Earth is roughly spherical, but it bulges towards the Equator, so it is roughly in the shape of an ellipsoid. This bulge is due to its rotation and is nearly consistent with an Earth in hydrostatic equilibrium. The detailed shape of the Earth, however, is also affected by the distribution of continents and ocean basins, and to some extent by the dynamics of the plates.

Structure of the Interior

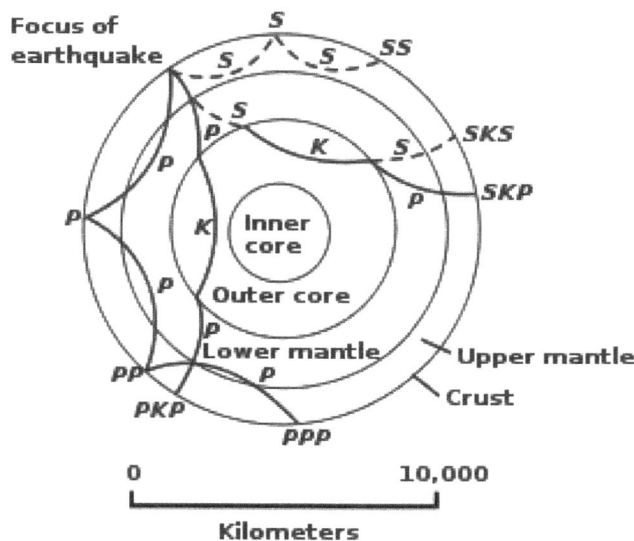

Seismic velocities and boundaries in the interior of the Earth sampled by seismic waves.

Evidence from seismology, heat flow at the surface, and mineral physics is combined with the Earth's mass and moment of inertia to infer models of the Earth's interior – its composition, density, temperature, pressure. For example, the Earth's mean specific gravity (5.515) is far higher than the typical specific gravity of rocks at the surface (2.7–3.3), implying that the deeper material is denser. This is also implied by its low moment of inertia ($0.33\,M\,R2$, compared to $0.4\,M\,R2$ for a sphere of constant density). However, some of the density increase is compression under the enormous pressures inside the Earth. The effect of pressure can be calculated using the Adams–Williamson equation. The conclusion is that pressure alone cannot account for the increase in density. Instead, we know that the Earth's core is composed of an alloy of iron and other minerals.

Reconstructions of seismic waves in the deep interior of the Earth show that there are no S-waves in the outer core. This indicates that the outer core is liquid, because liquids cannot support shear. The outer core is liquid, and the motion of this highly conductive fluid generates the Earth's field. The inner core, however, is solid because of the enormous pressure.

Reconstruction of seismic reflections in the deep interior indicate some major discontinuities in seismic velocities that demarcate the major zones of the Earth: inner core, outer core, mantle, lithosphere and crust. The mantle itself is divided into the upper mantle, transition zone, lower mantle and D'' layer. Between the crust and the mantle is the Mohorovičić discontinuity.

The seismic model of the Earth does not by itself determine the composition of the layers. For a complete model of the Earth, mineral physics is needed to interpret seismic velocities in terms of composition. The mineral properties are temperature-dependent, so the geotherm must also be determined. This requires physical theory for thermal conduction and convection and the heat contribution of radioactive elements. The main model for the radial structure of the interior of the Earth is the preliminary reference Earth model (PREM). Some parts of this model have been updated by recent findings in mineral physics and supplemented by seismic tomography. The mantle is mainly composed of silicates, and the boundaries between layers of the mantle are consistent with phase transitions.

The mantle acts as a solid for seismic waves, but under high pressures and temperatures it deforms so that over millions of years it acts like a liquid. This makes plate tectonics possible. Geodynamics is the study of the fluid flow in the mantle and core.

Magnetosphere

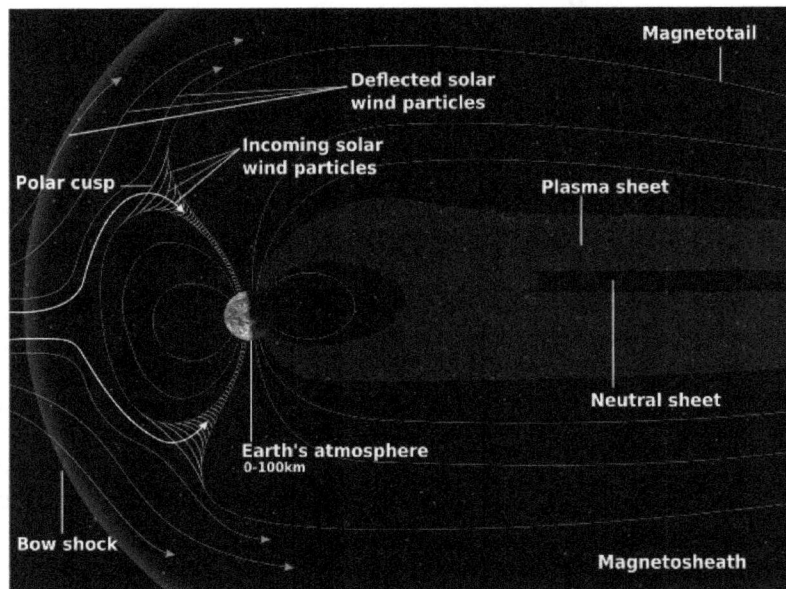

Schematic of Earth's magnetosphere. The solar wind flows from left to right.

If a planet's magnetic field is strong enough, its interaction with the solar wind forms a magnetosphere. Early space probes mapped out the gross dimensions of the Earth's magnetic field, which extends about 10 Earth radii towards the Sun. The solar wind, a stream of charged particles, streams out and around the terrestrial magnetic field, and continues behind the magnetic tail, hundreds of Earth radii downstream. Inside the magnetosphere, there are relatively dense regions of solar wind particles called the Van Allen radiation belts.

Methods

Geodesy

Geophysical measurements are generally at a particular time and place. Accurate measurements of position, along with earth deformation and gravity, are the province of geodesy. While geodesy and geophysics are separate fields, the two are so closely connected that many scientific organizations such as the American Geophysical Union, the Canadian Geophysical Union and the International Union of Geodesy and Geophysics encompass both.

Absolute positions are most frequently determined using the global positioning system (GPS). A three-dimensional position is calculated using messages from four or more visible satellites and referred to the 1980 Geodetic Reference System. An alternative, optical astronomy, combines astronomical coordinates and the local gravity vector to get geodetic coordinates. This method only provides the position in two coordinates and is more difficult to use than GPS. However, it is useful for measuring motions of the Earth such as nutation and Chandler wobble. Relative positions of two or more points can be determined using very-long-baseline interferometry.

Gravity measurements became part of geodesy because they were needed to related measurements at the surface of the Earth to the reference coordinate system. Gravity measurements on land can be made using gravimeters deployed either on the surface or in helicopter flyovers. Since the 1960s, the Earth's gravity field has been measured by analyzing the motion of satellites. Sea level can also be measured by satellites using radar altimetry, contributing to a more accurate geoid. In 2002, NASA launched the Gravity Recovery and Climate Experiment (GRACE), wherein two twin satellites map variations in Earth's gravity field by making measurements of the distance between the two satellites using GPS and a microwave ranging system. Gravity variations detected by GRACE include those caused by changes in ocean currents; runoff and ground water depletion; melting ice sheets and glaciers.

Space Probes

Space probes made it possible to collect data from not only the visible light region, but in other areas of the electromagnetic spectrum. The planets can be characterized by their force fields: gravity and their magnetic fields, which are studied through geophysics and space physics.

Measuring the changes in acceleration experienced by spacecraft as they orbit has allowed fine details of the gravity fields of the planets to be mapped. For example, in the 1970s, the gravity field disturbances above lunar maria were measured through lunar orbiters, which led to the discovery of concentrations of mass, mascons, beneath the Imbrium, Serenitatis, Crisium, Nectaris and Humorum basins.

History

Geophysics emerged as a separate discipline only in the 19th century, from the intersection of physical geography, geology, astronomy, meteorology, and physics. However, many geophysical phenomena – such as the Earth's magnetic field and earthquakes – have been investigated since the ancient era.

Ancient and Classical Eras

The magnetic compass existed in China back as far as the fourth century BC. It was used as much for feng shui as for navigation on land. It was not until good steel needles could be forged that compasses were used for navigation at sea; before that, they could not retain their magnetism long enough to be useful. The first mention of a compass in Europe was in 1190 AD.

Replica of Zhang Heng's seismoscope, possibly the first contribution to seismology.

In circa 240 BC, Eratosthenes of Cyrene deduced that the Earth was round and measured the circumference of the Earth, using trigonometry and the angle of the Sun at more than one latitude in Egypt. He developed a system of latitude and longitude.

Perhaps the earliest contribution to seismology was the invention of a seismoscope by the prolific inventor Zhang Heng in 132 AD. This instrument was designed to drop a bronze ball from the mouth of a dragon into the mouth of a toad. By looking at which of eight toads had the ball, one could determine the direction of the earthquake. It was 1571 years before the first design for a seismoscope was published in Europe, by Jean de la Hautefeuille. It was never built.

Beginnings of Modern Science

One of the publications that marked the beginning of modern science was William Gilbert's *De Magnete* (1600), a report of a series of meticulous experiments in magnetism. Gilbert deduced that compasses point north because the Earth itself is magnetic.

In 1687 Isaac Newton published his *Principia,* which not only laid the foundations for classical mechanics and gravitation but also explained a variety of geophysical phenomena such as the tides and the precession of the equinox.

The first seismometer, an instrument capable of keeping a continuous record of seismic activity, was built by James Forbes in 1844.

Geochemistry

Geochemistry is the science that uses the tools and principles of chemistry to explain the mechanisms behind major geological systems such as the Earth's crust and its oceans. The realm of geochemistry extends beyond the Earth, encompassing the entire Solar System and has made important contributions to the understanding of a number of processes including mantle convection, the formation of planets and the origins of granite and basalt.

History

The term *geochemistry* was first used by the Swiss-German chemist Christian Friedrich Schönbein in 1838. In his paper, Schönbein predicted the birth of a new field of study, stating:

"In a word, a comparative geochemistry ought to be launched, before geochemistry can become geology, and before the mystery of the genesis of our planets and their inorganic matter may be revealed."

The field began to be realised a short time after Schönbein's work, but his term - 'geochemistry' - was initially used neither by geologists nor chemists and there was much debate over which of the two sciences should be the dominant partner. There was little collaboration between geologists and chemists and the field of geochemistry remained small and unrecognised. In the late 19th Century a Swiss man by the name of Victor Goldschmidt was born, who later became known as the father of geochemistry. His paper, Geochemische Verteilungsgesetze der Elemente, on the distribution of elements in nature has been referred to as the start of geochemistry. During the early 20th Century, a number of geochemists produced work that began to popularise the field, including Frank Wigglesworth Clarke who had begun to investigate the abundances of various elements within the Earth and how the quantities were related to atomic weight. The composition of meteorites and their differences to terrestrial rocks was being investigated as early as 1850 and in 1901, Oliver C. Farrington hypothesised although there were differences, that the relative abundances should still be the same. This was the beginnings of the field of cosmochemistry and has contributed much of what we know about the formation of the Earth and the Solar System.

Subfields

Some subsets of geochemistry are:

1. Isotope geochemistry involves the determination of the relative and absolute concentrations of the elements and their isotopes in the earth and on earth's surface.

2. Examination of the distribution and movements of elements in different parts of the earth (crust, mantle, hydrosphere etc.) and in minerals with the goal to determine the underlying system of distribution and movement.

3. Cosmochemistry includes the analysis of the distribution of elements and their isotopes in the cosmos.

4. Biogeochemistry is the field of study focusing on the effect of life on the chemistry of the earth.

5. Organic geochemistry involves the study of the role of processes and compounds that are derived from living or once-living organisms.

6. Aqueous geochemistry studies the role of various elements in watersheds, including cop-

per, sulfur, mercury, and how elemental fluxes are exchanged through atmospheric-terrestrial-aquatic interactions.

7. Regional, environmental and exploration geochemistry includes applications to environmental, hydrological and mineral exploration studies.

8. Photogeochemistry is the study of light-induced chemical reactions that occur or may occur among natural components of the earth's surface.

Victor Goldschmidt is considered by most to be the father of modern geochemistry and the ideas of the subject were formed by him in a series of publications from 1922 under the title 'Geochemische Verteilungsgesetze der Elemente' (geochemical laws of distribution of the elements).

Chemical Characteristics

The more common rock constituents are nearly all oxides; chlorides, sulfides and fluorides are the only important exceptions to this and their total amount in any rock is usually much less than 1%. F. W. Clarke has calculated that a little more than 47% of the Earth's crust consists of oxygen. It occurs principally in combination as oxides, of which the chief are silica, alumina, iron oxides, and various carbonates (calcium carbonate, magnesium carbonate, sodium carbonate, and potassium carbonate). The silica functions principally as an acid, forming silicates, and all the commonest minerals of igneous rocks are of this nature. From a computation based on 1672 analyses of numerous kinds of rocks Clarke arrived at the following as the average percentage composition of the earths crust: $SiO_2=59.71$, $Al_2O_3=15.41$, $Fe_2O_3=2.63$, $FeO=3.52$, $MgO=4.36$, $CaO=4.90$, $Na_2O=3.55$, $K_2O=2.80$, $H_2O=1.52$, $TiO_2=0.60$, $P_2O_5=0.22$, (total 99.22%). All the other constituents occur only in very small quantities, usually much less than 1%.

These oxides combine in a haphazard way. For example, potash (potassium carbonate) and soda (sodium carbonate) combine to produce feldspars. In some cases they may take other forms, such as nepheline, leucite, and muscovite, but in the great majority of instances they are found as feldspar. Phosphoric acid with lime (calcium carbonate) forms apatite. Titanium dioxide with ferrous oxide gives rise to ilmenite. Part of the lime forms lime feldspar. Magnesium carbonate and iron oxides with silica crystallize as olivine or enstatite, or with alumina and lime form the complex ferro-magnesian silicates of which the pyroxenes, amphiboles, and biotites are the chief. Any excess of silica above what is required to neutralize the bases will separate out as quartz; excess of alumina crystallizes as corundum. These must be regarded only as general tendencies. It is possible, by rock analysis, to say approximately what minerals the rock contains, but there are numerous exceptions to any rule.

Mineral Constitution

Hence we may say that except in acid or siliceous rocks containing greater than 66% of silica are known as felsic rocks, and Quartz is not abundant. In basic rocks (containing 20% of silica or less) it is rare for them to contain as much silicon, these are referred to as mafic rocks. If Magnesium and Iron are above average while silica is low, olivine may be expected; where silica is present in greater quantity over ferro-magnesian minerals, such as augite, hornblende, enstatite or biotite, occur rather than olivine. Unless potash is high and silica relatively low, leucite will not be present, for leucite does not occur with free quartz. Nepheline, likewise, is usually found in rocks with much soda and comparatively little silica. With high alkalis, soda-bearing pyroxenes and amphiboles may be present. The lower the percentage of silica and alkali's, the greater is the prevalence of plagioclase feldspar as contracted with soda or potash feldspar. The earth crust is composed of 90% silicate minerals and their abundance in the earth is as follows; plagioclase feldspar (39%), Alkali feldspar (12%), quartz (12%), pyroxene (11%), amphiboles (5%), micas (5%), clay minerals (5%), after this the remaining silicate minerals make up another 3% of the earths crust. Only 8% of the earth is composed of non silicate minerals such as Carbonate, Oxides, and Sulfides.

The other determining factor, namely the physical conditions attending consolidation, plays on the whole a smaller part, yet is by no means negligible, as a few instances will prove. Certain minerals are practically confined to deep-seated intrusive rocks, e.g., microcline, muscovite, diallage. Leucite is very rare in plutonic masses; many minerals have special peculiarities in microscopic character according to whether they crystallized in depth or near the surface, e.g., hypersthene, orthoclase, quartz. There are some curious instances of rocks having the same chemical composition, but consisting of entirely different minerals, e.g., the hornblendite of Gran, in Norway, which contains only hornblende, has the same composition as some of the camptonites of the same locality that contain feldspar and hornblende of a different variety. In this connection we may repeat what has been said above about the corrosion of porphyritic minerals in igneous rocks. In rhyolites and trachytes, early crystals of hornblende and biotite may be found in great numbers partially converted into augite and magnetite. Hornblende and biotite were stable under the pressures and other conditions below the surface, but unstable at higher levels. In the ground-mass of these rocks, augite is almost universally present. But the plutonic representatives of the same magma, granite and syenite contain biotite and hornblende far more commonly than augite.

Felsic, Intermediate and Mafic Igneous Rocks

Those rocks that contain the most silica, and on crystallizing yield free quartz, form a group generally designated the "felsic" rocks. Those again that contain least silica and most magnesia and iron, so that quartz is absent while olivine is usually abundant, form the "mafic" group. The "intermediate" rocks include those characterized by the general absence of both quartz and olivine. An important subdivision of these contains a very high percentage of alkalis, especially soda, and consequently has minerals such as nepheline and leucite not common in other rocks. It is often separated from the others as the "alkali" or "soda" rocks, and there is a corresponding series of mafic rocks. Lastly a small sub-group rich in olivine and without feldspar has been called the "ultramafic" rocks. They have very low percentages of silica but much iron and magnesia.

Except these last, practically all rocks contain felspars or feldspathoid minerals. In the acid rocks the common feldspars are orthoclase, perthite, microcline, and oligoclase—all having much silica and alkalis. In the mafic rocks labradorite, anorthite and bytownite prevail, being rich in lime and poor in silica, potash and soda. Augite is the most common ferro-magnesian in mafic rocks, but biotite and hornblende are on the whole more frequent in felsic rocks.

Most Common Minerals	Acid	Intermediate		Mafic	Ultramafic
	Quartz Orthoclase (and Oligoclase), Mica, Hornblende, Augite	Little or no Quartz: Orthoclase hornblende, Augite, Biotite	Little or no Quartz: Plagioclase Hornblende, Augite, Biotite	No Quartz Plagioclase Augite, Olivine	No Felspar Augite, Hornblende, Olivine
Plutonic or Abyssal type	Granite	Syenite	Diorite	Gabbro	Peridotite
Intrusive or Hypabyssal type	Quartz-porphyry	Orthoclase-porphyry	Porphyrite	Dolerite	Picrite
Lavas or Effusive type	Rhyolite, Obsidian	Trachyte	Andesite	Basalt	Limburgite

Rocks that contain leucite or nepheline, either partly or a wholly replacing felspar, are not included in this table. They are essentially of intermediate or of mafic character. We might in consequence regard them as varieties of syenite, diorite, gabbro, etc., in which feldspathoid minerals occur, and indeed there are many transitions between syenites of ordinary type and nepheline — or leucite — syenite, and between gabbro or dolerite and theralite or essexite. But, as many minerals develop in these "alkali" rocks that are uncommon elsewhere, it is convenient in a purely formal classification like that outlined here to treat the whole assemblage as a distinct series.

Nepheline and Leucite-bearing Rocks		
Most Common Minerals	Alkali Feldspar, Nepheline or Leucite, Augite, Hornblend, Biotite	Soda Lime Feldspar, Nepheline or Leucite, Augite, Hornblende (Olivine)	Nepheline or Leucite, Augite, Hornblende, Olivine
Plutonic type	Nepheline-syenite, Leucite-syenite, Nepheline-porphyry	Essexite and Theralite	Ijolite and Missourite
Effusive type or Lavas	Phonolite, Leucitophyre	Tephrite and Basanite	Nepheline-basalt, Leucite-basalt

This classification is based essentially on the mineralogical constitution of the igneous rocks. Any chemical distinctions between the different groups, though implied, are relegated to a subordinate position. It is admittedly artificial but it has grown up with the growth of the science and is still adopted as the basis on which more minute subdivisions are erected. The subdivisions are by no means of equal value. The syenites, for example, and the peridotites, are far less important than the granites, diorites and gabbros. Moreover, the effusive andesites do not always correspond to the plutonic diorites but partly also to the gabbros. As the different kinds of rock, regarded as aggregates of minerals, pass gradually into one another, transitional types are very common and are often so important as to receive special names. The quartz-syenites and nordmarkites may be interposed between granite and syenite, the tonalites and adamellites between granite and diorite, the monzoaites between syenite and diorite, norites and hyperites between diorite and gabbro, and so on.

Geochemistry of Trace Metals in the Ocean

Trace metals readily form complexes with major ions in the ocean, including hydroxide, carbonate, and chloride and their chemical speciation changes depending on whether the environment is oxidized or reduced. Benjamin (2002) defines complexes of metals with more than one type of ligand, other than water, as mixed-ligand-complexes. In some cases, a ligand contains more than one *donor* atom, forming very strong complexes, also called chelates (the ligand is the chelator). One of the most common chelators is EDTA (ethylenediaminetetraacetic acid), which can replace six molecules of water and form strong bonds with metals that have a plus two charge. With stronger complexation, lower activity of the free metal ion is observed. One consequence of the lower reactivity of complexed metals compared to the same concentration of free metal is that the chelation tends to stabilize metals in the aqueous solution instead of in solids.

Concentrations of the trace metals cadmium, copper, molybdenum, manganese, rhenium, uranium and vanadium in sediments record the redox history of the oceans. Within aquatic environments, cadmium(II) can either be in the form $CdCl+(aq)$ in oxic waters or $CdS(s)$ in a reduced environment. Thus higher concentrations of Cd in marine sediments may indicate low redox potential conditions in the past. For copper(II), a prevalent form is $CuCl+(aq)$ within oxic environments and $CuS(s)$ and Cu_2S within reduced environments. The reduced seawater environment leads to two possible oxidation states of copper, Cu(I) and Cu(II). Molybdenum is present as the Mo(VI) oxidation state as $MoO_4{}^{2-}(aq)$ in oxic environments. Mo(V) and Mo(IV) are present in reduced environments in the forms $MoO_2{}^+(aq)$ and $MoS_2(s)$. Rhenium is present as the Re(VII) oxidation state as $ReO_4{}^-$ within oxic conditions, but is reduced to Re(IV) which may form ReO_2 or ReS_2. Uranium is in oxidation state VI in $UO_2(CO_3)_3{}^{4-}(aq)$ and is found in the reduced form $UO_2(s)$. Vanadium is in several forms in oxidation state V(V); $HVO_4{}^{2-}$ and $H_2VO_4{}^-$. Its reduced forms can include $VO_2{}^+$, $VO(OH)_3{}^-$, and $V(OH)_3$. These relative dominance of these species depends on pH.

In the water column of the ocean or deep lakes, vertical profiles of dissolved trace metals are characterized as following *conservative–type, nutrient–type, or scavenged–type* distributions. Across these three distributions, trace metals have different residence times and are used to varying extents by planktonic microorganisms. Trace metals with conservative-type distributions have high concentrations relative to their biological use. One example of a trace metal with a conservative-type distribution is molybdenum. It has a residence time within the oceans of around 8×10^5 years and is generally present as the molybdate anion ($MoO_4{}^{2-}$). Molybdenum interacts weakly with particles and displays an almost uniform vertical profile in the ocean. Relative to the abundance of molybdenum in the ocean, the amount required as a metal cofactor for enzymes in marine phytoplankton is negligible.

Trace metals with nutrient-type distributions are strongly associated with the internal cycles of particulate organic matter, especially the assimilation by plankton. The lowest dissolved concentrations of these metals are at the surface of the ocean, where they are assimilated by plankton. As dissolution and decomposition occur at greater depths, concentrations of these trace metals increase. Residence times of these metals, such as zinc, are several thousand to one hundred thousand years. Finally, an example of a scavenged-type trace metal is aluminium, which has strong interactions with particles as well as a short residence time in the ocean. The residence times of scavenged-type trace metals are around 100 to 1000 years. The concentrations of these metals are highest around bottom sediments, hydrothermal vents, and rivers. For aluminium, atmospheric dust provides the greatest source of external inputs into the ocean.

Iron and copper show hybrid distributions in the ocean. They are influenced by recycling and intense scavenging. Iron is a limiting nutrient in vast areas of the oceans, and is found in high abundance along with manganese near hydrothermal vents. Here, many iron precipitates are found, mostly in the forms of iron sulfides and oxidized iron oxyhydroxide compounds. Concentrations of iron near hydrothermal vents can be up to one million times the concentrations found in the open ocean.

Using electrochemical techniques, it is possible to show that bioactive trace metals (zinc, cobalt, cadmium, iron and copper) are bound by organic ligands in surface seawater. These ligand complexes serve to lower the bioavailability of trace metals within the ocean. For example, copper, which may be toxic to open ocean phytoplankton and bacteria, can form organic complexes. The formation of these complexes reduces the concentrations of bioavailable inorganic complexes of copper that could be toxic to sea life at high concentrations. Unlike copper, zinc toxicity in marine phytoplankton is low and there is no advantage to increasing the organic binding of Zn^{2+}. In high nutrient-low chlorophyll regions, iron is the limiting nutrient, with the dominant species being strong organic complexes of Fe(III).

Geobiology

The colorful microbial mats of Grand Prismatic Spring in Yellowstone National Park, USA. The orange mats are composed of Chloroflexi, Cyanobacteria, and other organisms that thrive in the 70 °C water. Geobiologists often study extreme environments like this because they are home to extremophilic organisms. It has been hypothesized that these environments may be representative of early Earth.

Geobiology is a field of scientific research that explores the interactions between the physical Earth and the biosphere. It is a relatively young field, and its borders are fluid. There is considerable overlap with the fields of ecology, evolutionary biology, microbiology, paleontology, and particularly biogeochemistry. Geobiology applies the principles and methods of biology and geology to the study of the ancient history of the co-evolution of life and Earth as well as the role of life in the modern world. Geobiologic studies tend to be focused on microorganisms, and on the role that life plays in altering the chemical and physical environment of the lithosphere, atmosphere, hydrosphere and/or cryosphere. It differs from biogeochemistry in that the focus is on processes and organisms over space and time rather than on global chemical cycles.

Geobiological research synthesizes the geologic record with modern biologic studies. It deals with process - how organisms affect the Earth and vice versa - as well as history - how the Earth and life have changed together. Much research is grounded in the search for fundamental understanding, but geobiology can also be applied, as in the case of microbes that clean up oil spills.

Geobiology employs molecular biology, environmental microbiology, chemical analyses, and the geologic record to investigate the evolutionary interconnectedness of life and Earth. It attempts to understand how the Earth has changed since the origin of life and what it might have been like along the way. Some definitions of geobiology even push the boundaries of this time frame - to understanding the origin of life and to the role that man has played and will continue to play in shaping the Earth in the Anthropocene.

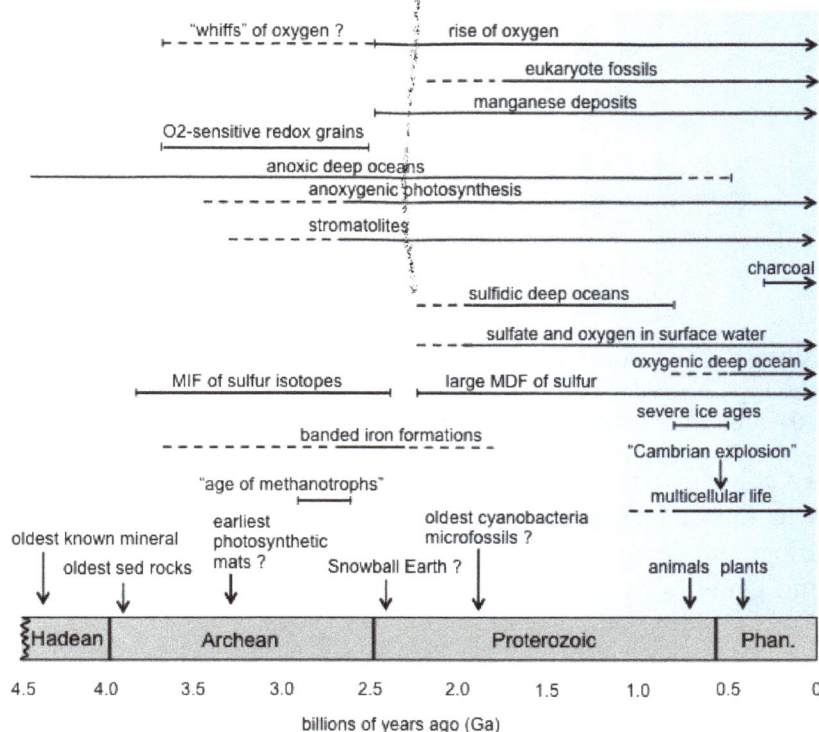

The geologic timescale overlain with major geobiologic events and occurrences. The oxygenation of the atmosphere is shown in blue starting 2.4 Ga, although the exact dating of the Great Oxygenation Event is debated.

History

The term geobiology was coined by Lourens Baas Becking in 1934. In his words, geobiology "is an attempt to describe the relationship between organisms and the Earth," for "the organism is part of the Earth and its lot is interwoven with that of the Earth." Baas Becking's definition of geobiology was born of a desire to unify environmental biology with laboratory biology. The way he practiced it aligns closely with modern environmental microbial ecology, though his definition remains applicable to all of geobiology. In his book, Geobiology, Bass Becking stated that he had no intention of inventing a new field of study.

Baas Becking's understanding of geobiology was heavily influenced by his predecessors, including

Martinus Beyerinck, his teacher from the Dutch School of Microbiology. Others included Vladimir Vernadsky, who argued that life changes the surface environment of Earth in The Biosphere, his 1926 book, and Sergei Vinogradsky, famous for discovering lithotrophic bacteria.

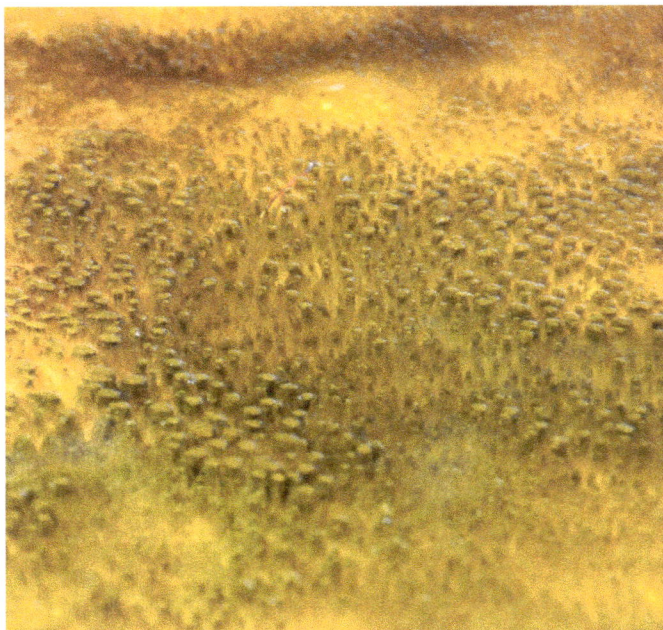

A microbial mat in White Creek, Yellowstone National Park, USA. Note the conical microstructure of the bacterial communities. These are hypothesized to be a living analogue of ancient fossil stromatolites. Each cone has an oxygen gas bubble on top, the product of oxygenic photosynthesis by cyanobacteria in the multi-species microbial mats.

The first laboratory officially dedicated to the study of geobiology was the Baas Becking Geobiological Laboratory in Australia, which opened its doors in 1965. However, it took another 40 or so years for geobiology to become a firmly rooted scientific discipline, thanks in part to advances in geochemistry and genetics that enabled scientists to begin to synthesize the study of life and planet.

In the 1930s, Alfred Treibs discovered chlorophyll-like porphyrins in petroleum, confirming its biological origin, thereby founding organic geochemistry and establishing the notion of biomarkers, a critical aspect of geobiology. But several decades passed before the tools were available to begin to search in earnest for chemical marks of life in the rocks. In the 1970s and 80s, scientists like Geoffrey Eglington and Roger Summons began to find lipid biomarkers in the rock record using equipment like GCMS.

On the biology side of things, in 1977, Carl Woese and George Fox published a phylogeny of life on Earth, including a new domain - the Archaea. And in the 1990s, genetics and genomics studies became possible, broadening the scope of investigation of the interaction of life and planet.

Today, geobiology has its own journals, such as *Geobiology, established in 2003,* and *Biogeosciences, established in 2004,* as well as recognition at major scientific conferences. It got its own Gordon Research Conference in 2011, a number of geobiology textbooks have been published, and many universities around the world offer degree programs in geobiology.

Major Geobiological Events

Graphical Anthropocentric representation of Earth's history as a spiral

Perhaps the most profound geobiological event is the introduction of oxygen into the atmosphere by photosynthetic bacteria. This oxygenation of Earth's primoidial atmosphere (the so-called oxygen catastrophe or Great Oxygenation Event) and the oxygenation of the oceans altered surface biogeochemical cycles and the types of organisms that have been evolutionarily selected for.

A subsequent major change was the advent of multicellularity. The presence of oxygen allowed eukaryotes and, later, multicellular life to evolve.

More anthropocentric geobiologic events include the origin of animals and the establishment of terrestrial plant life, which affected continental erosion and nutrient cycling, and likely changed the types of rivers observed, allowing channelization of what were previously predominantly braided rivers.

More subtle geobiological events include the role of termites in overturning sediments, coral reefs in depositing calcium carbonate and breaking waves, sponges in absorbing dissolved marine silica, the role of dinosaurs in breaching river levees and promoting flooding, and the role of large mammal dung in distributing nutrients.

Important Concepts

Geobiology is founded upon a few core concepts that unite the study of Earth and life. While there are many aspects of studying past and present interactions between life and Earth that are unclear, several important ideas and concepts provide a basis of knowledge in geobiology that serve as a platform for posing researchable questions, including the evolution of life and planet and the co-evolution of the two, genetics - from both a historical and functional standpoint, the metabolic diversity of all life, the sedimentological preservation of past life, and the origin of life.

Co-Evolution of Life and Earth

A core concept in geobiology is that life changes over time through evolution. The theory of evolution postulates that unique populations of organisms or species arose from genetic modifications in the ancestral population which were passed down by drift and natural selection.

Along with standard biological evolution, life and planet co-evolve. Since the best adaptations are those that suit the ecological niche that the organism lives in, the physical and chemical characteristics of the environment drive the evolution of life by natural selection, but the opposite can also be true: with every advent of evolution, the environment changes.

A classic example of co-evolution is the evolution of oxygen-producing photosynthetic cyanobacteria which oxygenated Earth's Archean atmosphere. The ancestors of cyanobacteria began using water as an electron source to harness the energy of the sun and expelling oxygen before or during the early Paleoproterozoic. During this time, around 2.4 to 2.1 billion years ago, geologic data suggests that atmospheric oxygen began to rise in what is termed the Great Oxygenation Event (GOE). It is unclear for how long cyanobacteria had been doing oxygenic photosynthesis before the GOE. Some evidence suggests there were geochemical "buffers" or sinks suppressing the rise of oxygen such as volcanism though cyanobacteria may have been around producing it before the GOE. Other evidence indicates that the rise of oxygenic photosynthesis was coincident with the GOE.

Banded iron formation (BIF), Hammersley Formation, Western Australia

The presence of oxygen on Earth from its first production by cyanobacteria to the GOE and through today has drastically impacted the course of evolution of life and planet. It may have triggered the formation of oxidized minerals and the disappearance of oxidizable minerals like pyrite from ancient stream beds. The presence of banded-iron formations (BIFs) have been interpreted as a clue for the rise of oxygen since small amounts of oxygen could have reacted with reduced ferrous iron ($Fe(II)$) in the oceans, resulting in the deposition of sediments containing $Fe(III)$ oxide in places like Western Australia. However, any oxidizing environment, including that provided by microbes such as the iron-oxidizing photoautotroph *Rhodopseudomonas palustris,* can trigger iron oxide formation and thus BIF deposition. Other mechanisms include oxidation by UV light. Indeed, BIFs occur across large swaths of Earth's history and may not correlate with only one event.

Other changes correlated with the rise of oxygen include the appearance of rust-red ancient pa-leosols, different isotope fractionation of elements such as sulfur, and global glaciations and Snow-ball Earth events, perhaps caused by the oxidation of methane by oxygen, not to mention an over-haul of the types of organisms and metabolisms on Earth. Whereas organisms prior to the rise of oxygen were likely poisoned by oxygen gas as many anaerobes are today, those that evolved ways to harness the electron-accepting and energy-giving power of oxygen were poised to thrive and colonize the aerobic environment.

Modern, living stromatolites in Shark Bay, Australia. Shark Bay is one of the few places in the world where stromato-lites can be seen today, though they were likely common in ancient shallow seas before the rise of metazoan predators.

The Earth has Changed

Earth has not remained the same since its planetary formation 4.5 billion years ago. Continents have formed, broken up, and collided, offering new opportunities for and barriers to the dispersal of life. The redox state of the atmosphere and the oceans has changed, as indicated by isotope data. Fluctuating quantities of inorganic compounds such as carbon dioxide, nitrogen, methane, and oxygen have been driven by life evolving new biological metabolisms to make these chemicals and have driven the evolution of new metabolisms to use those chemicals. Earth acquired a magnetic field about 3.4 Ga that has undergone a series of geomagnetic reversals on the order of millions of years. The surface temperature is in constant fluctuation, falling in glaciations and Snowball Earth events due to ice-albedo feedback, rising and melting due to volcanic outgassing, and stabilizing due to silicate weathering feedback.

And the Earth is not the only one that changed - the luminosity of the sun has increased over time. Because rocks record a history of relatively constant temperatures since Earth's beginnings, there must have been more greenhouse gasses to keep the temperatures up in the Archean when the sun was younger and fainter. All these major differences in the environment of the Earth placed very different constraints on the evolution of life throughout our planet's history. Moreover, more subtle changes in the habitat of life are always occurring, shaping the organisms and traces that we observe today and in the rock record.

Genes Encode Geobiological Function and History

The genetic code is key to observing the history of evolution and understanding the capabilities of organisms. Genes are the basic unit of inheritance and function and, as such, they are the basic unit of evolution and the means behind metabolism.

Phylogeny Predicts Evolutionary History

Phylogenetic Tree of Life

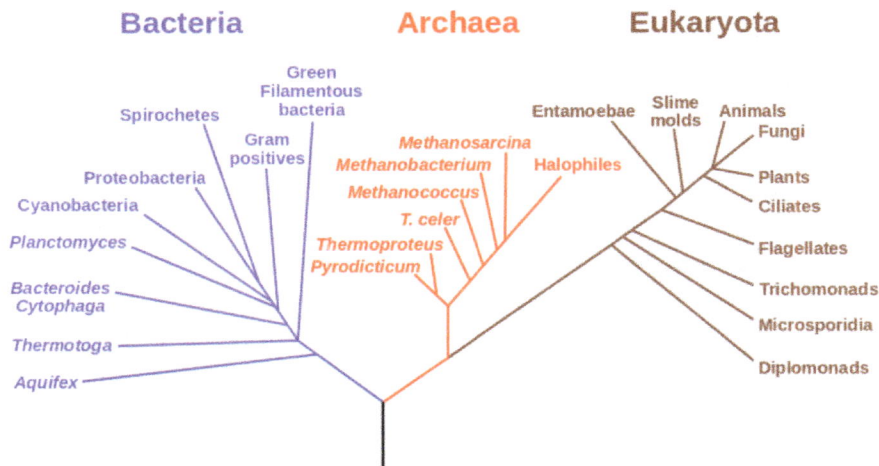

A phylogenetic tree of living things, based on rRNA data and proposed by Carl Woese, showing the separation of bacteria, archaea, and eukaryotes and linking the three branches of living organisms to the LUCA (the black trunk at the bottom of the tree).

Phylogeny takes genetic sequences from living organisms and compares them to each other to reveal evolutionary relationships, much like a family tree reveals how individuals are connected to their distant cousins. It allows us to decipher modern relationships and infer how evolution happened in the past.

Phylogeny can give some sense of history when combined with a little bit more information. Each difference in the DNA indicates divergence between one species and another. This divergence, whether via drift or natural selection, is representative of some lapse of time. Comparing DNA sequences alone gives a record of the history of evolution with an arbitrary measure of phylogenetic distance "dating" that last common ancestor. However, if information about the rate of genetic mutation is available or geologic markers are present to calibrate evolutionary divergence (i.e. fossils), we have a timeline of evolution. From there, with an idea about other contemporaneous changes in life and environment, we can begin to speculate why certain evolutionary paths might have been selected for.

Genes Encode Metabolism

Molecular biology allows scientists to understand a gene's function using microbial culturing and mutagenesis. Searching for similar genes in other organisms and in metagenomic and metatranscriptomic data allows us to understand what processes could be relevant and important in a given ecosystem, providing insight into the biogeochemical cycles in that environment.

For example, an intriguing problem in geobiology is the role of organisms in the global cycling of methane. Genetics has revealed that the methane monooxygenase gene (*pmo*) is used for oxidizing methane and is present in all aerobic methane-oxidizers, or methanotrophs. The presence of DNA sequences of the *pmo* gene in the environment can be used as a proxy for methanotrophy. A more generalizable tool is the 16S ribosomal RNA gene, which is found in bacteria and archaea. This gene evolves very slowly over time and is not usually horizontally transferred, and so it is often used to distinguish different taxonomic units of organisms in the environment. In this way, genes are clues to organismal metabolism and identity. Genetics enables us to ask 'who is there?' and 'what are they doing?' This approach is called metagenomics.

3.4 billion year-old stromatolites from the Warrawoona Group, Western Australia. While the origin of Precambrian stromatolites is a heavily debated topic in geobiology, stromatolites from Warrawoona are hypothesized to have been formed by ancient communities of microbes.

Metabolic Diversity Influences the Environment

Life harnesses chemical reactions to generate energy, perform biosynthesis, and eliminate waste. Different organisms use very different metabolic approaches to meet these basic needs. While animals such as ourselves are limited to aerobic respiration, other organisms can "breathe" sulfate (SO_4^{2-}), nitrate (NO_3^-), ferric iron (Fe(III)), and uranium (U(VI)), or live off energy from fermentation. Some organisms, like plants, are autotrophs, meaning that they can fix carbon dioxide for biosynthesis. Plants are photoautotrophs, in that they use the energy of light to fix carbon. Microorganisms employ oxygenic and anoxygenic photoautotrophy, as well as chemoautotrophy. Microbial communities can coordinate in syntrophic metabolisms to shift reaction kinetics in their favor. Many organisms can perform multiple metabolisms to achieve the same end goal; these are called mixotrophs.

Biotic metabolism is directly tied to the global cycling of elements and compounds on Earth. The geochemical environment fuels life, which then produces different molecules that go into the external environment. (This is directly relevant to biogeochemistry.) In addition, biochemical reactions are catalyzed by enzymes which sometimes prefer one isotope over others. For example, oxygenic photosynthesis is catalyzed by RuBisCO, which prefers carbon-12 over carbon-13, resulting in carbon isotope fractionation in the rock record.

"Giant" ooids of the Johnnie Formation in the Death Valley area, California, USA. Ooids are near-spheroidal calcium carbonate grains that accumulate around a central nucleus and can be sedimented to form oolite like this. Microbes can mediate the formation of ooids.

Sedimentary Rocks Tell a Story

Sedimentary rocks preserve remnants of the history of life on Earth in the form of fossils, biomarkers, isotopes, and other traces. The rock record is far from perfect, and the preservation of biosignatures is a rare occurrence. Understanding what factors determine the extent of preservation and the meaning behind what is preserved are important components to detangling the ancient history of the co-evolution of life and Earth. The sedimentary record allows scientists to observe changes in life and Earth in composition over time and sometimes even date major transitions, like extinction events.

Some classic examples of geobiology in the sedimentary record include stromatolites and banded-iron formations. The role of life in the origin of both of these is a heavily debated topic.

Life is Fundamentally Chemistry

The first life arose from abiotic chemical reactions. When this happened, how it happened, and even what planet it happened on are uncertain. However, life follows the rules of and arose from lifeless chemistry and physics. It is constrained by principles such as thermodynamics. This is an important concept in the field because it is represents the epitome of the interconnectedness, if not sameness, of life and Earth.

While often delegated to the field of astrobiology, attempts to understand how and when life arose are relevant to geobiology as well. The first major strides towards understanding the "how" came with the Miller-Urey experiment, when amino acids formed out of a simulated "primordial soup". Another theory is that life originated in a system much like the hydrothermal vents at mid-oceanic spreading centers. In the Fischer-Tropsch synthesis, a variety of hydrocarbons form under vent-like conditions. Other ideas include the "RNA World" hypothesis, which postulates that the first biologic molecule was RNA and the idea that life originated elsewhere in the solar system and was brought to Earth, perhaps via a meteorite.

Methodology

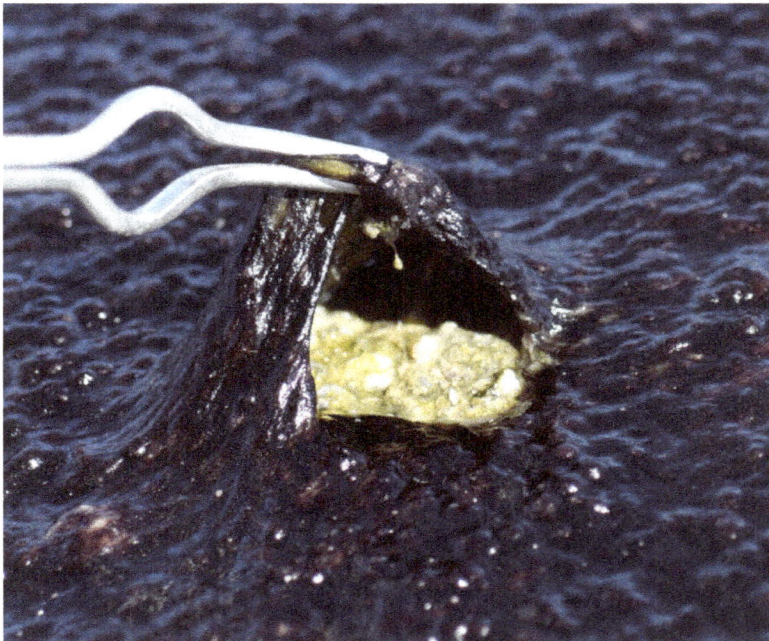

A microbial mat growing on acidic soil in Norris Geyser basin, Yellowstone National Park, USA. The black top serves as a sort of sunscreen, and when you look underneath you see the green cyanobacteria.

While geobiology is a diverse and varied field, encompassing ideas and techniques from a wide range of disciplines, there are a number of important methods that are key to the study of the interaction of life and Earth that are highlighted here.

1. *Laboratory culturing of microbes is used to characterize the metabolism and lifestyle of organisms of interest.*

2. *Gene sequencing allows scientists to study the relationships between extant organisms using phylogenetics.*

3. *Experimental genetic manipulation or* mutagenesis is used to determine the function of genes in living organisms.

4. *Microscopy is used to visualize the microbial world. Microscope work ranges from environmental observation to quantitative studies with* DNA probes to high-definition visualization of the microbe-mineral interface by electron microscope (EM).

5. *Isotope tracers can be used to track biochemical reactions to understand microbial metabolism.*

6. *Isotope natural abundance in rocks can be measured to look for* isotopic fractionation that is consistent with biologic origin.

7. Detailed *environmental characterization is important to understanding what about a habitat might be driving life's evolution and, in turn, how life might be changing that niche. It includes and is not limited to, temperature, light, pH, salinity, concentration of specific molecules like oxygen, and the biologic community.*

8. *Sedimentology and stratigraphy are used to read the rocks. The rock record stores a history of geobiologic processes in sediments which can be unearthed through an understanding of* deposition, sedimentation, compaction, diagenesis, and deformation.

9. The search for and study of fossils, while often delegated to the separate field of *paleontology, is important in geobiology, though the scale of fossils is typically smaller* (micropaleontology).

10. The biochemical analysis of *biomarkers, which are fossilized or modern molecules that are indicative of the presence of a certain group of organisms or metabolism, is used to answer the evidence for life and metabolic diversity questions.*

11. *Paleomagnetics is the study of the planet's ancient magnetic field. It is significant to understanding* magnetofossils, biomineralization, and global ecosystem changes.

Sub-Disciplines and Related Fields

As its name suggests, geobiology is closely related to many other fields of study, and does not have clearly defined boundaries or perfect agreement on what exactly they comprise. Some practitioners take a very broad view of its boundaries, encompassing many older, more established fields such as biogeochemistry, paleontology, and microbial ecology. Others take a more narrow view, assigning it to emerging research that falls between these existing fields, such as with geomicrobiology. The following list includes both those that are clearly a part of geobiology, e.g. geomicrobiology, as well as those that share scientific interests but have not historically been considered a sub-discipline of geobiology, e.g. paleontology.

Astrobiology

Astrobiology is an interdisciplinary field that uses a combination of geobiological and planetary science data to establish a context for the search for life on other planets. The origin of life from non-living chemistry and geology, or abiogenesis, is a major topic in astrobiology. Even though it is fundamentally an earth-bound concern, and therefore of great geobiological interest, getting at the origin of life necessitates considering what life requires, what, if anything, is special about Earth, what might have changed to allow life to blossom, what constitutes evidence for life, and even what constitutes life itself. These are the same questions that scientists might ask when searching for alien life. In addition, astrobiologists research the possibility of life based on other metabolisms and elements, the survivability of Earth's organisms on other planets or spacecrafts, planetary and solar system evolution, and space geochemistry.

Biogeochemistry

Biogeochemistry is a systems science that synthesizes the study of biological, geological, and chemical processes to understand the reactions and composition of the natural environment. It is concerned primarily with global elemental cycles, such as that of nitrogen and carbon. The father of biogeochemistry was James Lovelock, whose "Gaia hypothesis" proposed that Earth's biological, chemical, and geologic systems interact to stabilize the conditions on Earth that support life.

Geobiochemistry

Stromatolites in the Green River Shale, Wyoming, USA, dating to the Eocene.

Geobiochemistry is similar to biogeochemistry, but differs by placing emphasis on the effects of geology on the development of life's biochemical processes, as distinct from the role of life on Earth's cycles. Its primary goal is to link biological changes, encompassing evolutionary modifications of genes and changes in the expression of genes and proteins, to changes in the temperature, pressure, and composition of geochemical processes to understand when and how metabolism evolved. Geobiochemistry is founded on the notion that life is a planetary response because metabolic catalysis enables the release of energy trapped by a cooling planet.

Environmental Microbiology

Microbiology is a broad scientific discipline pertaining to the study of that life which is best viewed under a microscope. It encompasses several fields that are of direct relevance to geobiology, and the tools of microbiology all pertain to geobiology. Environmental microbiology is especially entangled in geobiology since it seeks an understanding of the actual organisms and processes that are relevant in nature, as opposed to the traditional lab-based approach to microbiology. Microbial ecology is similar, but tend to focus more on lab studies and the relationships between organisms within a community, as well as within the ecosystem of their chemical and geological physical environment. Both rely on techniques such as sample collection from diverse environments, metagenomics, DNA sequencing, and statistics.

Geomicrobiology and Microbial Geochemistry

A vertical cross section of a microbial mat containing different organisms that perform different metabolisms. The green are presumably cyanobacteria, and teepee-like microstructures are visible on the surface.

Geomicrobiology traditionally studies the interactions between microbes and minerals. While it is generally reliant on the tools of microbiology, microbial geochemistry uses geological and chemical methods to approach the same topic from the perspective of the rocks. Geomicrobiology and microbial geochemistry (GMG) is a relatively new interdisciplinary field that more broadly takes on the relationship between microbes, Earth, and environmental systems. Billed as a subset of both geobiology and geochemistry, GMG seeks to understand elemental biogeochemical cycles and the evolution of life on Earth. Specifically, it asks questions about where microbes live, their local and global abundance, their structural and functional biochemistry, how they have evolved, biomineralization, and their preservation potential and presence in the rock record. In many ways, GMG appears to be equivalent to geobiology, but differs in scope: geobiology focuses on the role of all life, while GMG is strictly microbial. Regardless, it is these tiniest creatures that dominated to history of life integrated over time and seem to have had the most far-reaching effects.

Molecular Geomicrobiology

Molecular geomicrobiology takes a mechanistic approach to understanding biological processes that are geologically relevant. It can be at the level of DNA, protein, lipids, or any metabolite.

Organic Geochemistry

Organic geochemistry is the study of organic molecules that appear in the fossil record in sedimentary rocks. Research in this field concerns molecular fossils that are often lipid biomarkers. Molecules like sterols and hopanoids, membrane lipids found in eukaryotes and bacteria, respectively, can be preserved in the rock record on billion-year timescales. Following the death of the organism they came from and sedimentation, they undergo a process called diagenesis whereby many of the specific functional groups from the lipids are lost, but the hydrocarbon skeleton remains intact. These fossilized lipids are called steranes and hopanes, respectively. There are also other types of molecular fossils, like porphyrins, the discovery of which in petroleum by Alfred E. Treibs actually led to the invention of the field. Other aspects of geochemistry that are also pertinent to geobiology include isotope geochemistry, in which scientists search for isotope fractionation in the rock record, and the chemical analysis of biominerals, such as magnetite or microbially-precipitated gold.

Ediacaran fossils from Mistaken Point, Newfoundland. Ediacaran biota originated during the Ediacaran Period and are unlike most animals around today.

Paleontology

Perhaps the oldest of the bunch, paleontology is the study of fossils. It involves the discovery, excavation, dating, and paleoecological understanding of any type of fossil, microbial or dinosaur, trace or body fossil. Micropaleontology is particularly relevant to geobiology. Putative bacterial microfossils and ancient stromatolites are used as evidence for the rise of metabolisms such as oxygenic photosynthesis. The search for molecular fossils, such as lipid biomarkers like steranes and hopanes, has also played an important role in geobiology and organic geochemistry. Relevant sub-disciples include paleoecology and paleobiogeoraphy.

Biogeography

Biogeography is the study of the geographic distribution of life through time. It can look at the present distribution of organisms across continents or between microniches, or the distribution of organisms through time, or in the past, which is called paleobiogeography.

Evolutionary Biology

Evolutionary biology is the study of the evolutionary processes that have shaped the diversity of life on Earth. It incorporates genetics, ecology, biogeography, and paleontology to analyze topics including natural selection, variance, adaptation, divergence, genetic drift, and speciation.

Geoinformatics

Geoinformatics is the science and the technology which develops and uses information science infrastructure to address the problems of geography, cartography, geosciences and related branches of science and engineering.

Overview

Geoinformatics has been described as "the science and technology dealing with the structure and character of spatial information, its capture, its classification and qualification, its storage, processing, portrayal and dissemination, including the infrastructure necessary to secure optimal use of this information" or "the art, science or technology dealing with the acquisition, storage, processing production, presentation and dissemination of geoinformation".

Geomatics is a similarly used term which encompasses geoinformatics, but geomatics focuses more so on surveying. Geoinformatics has at its core the technologies supporting the processes of acquiring, analyzing and visualizing spatial data. Both geomatics and geoinformatics include and rely heavily upon the theory and practical implications of geodesy.

Geography and earth science increasingly rely on digital spatial data acquired from remotely sensed images analyzed by geographical information systems (GIS) and visualized on paper or the computer screen.

Geoinformatics combines geospatial analysis and modeling, development of geospatial databases, information systems design, human-computer interaction and both wired and wireless networking technologies. Geoinformatics uses geocomputation and geovisualization for analyzing geoinformation.

Branches of geoinformatics include:

Cartography

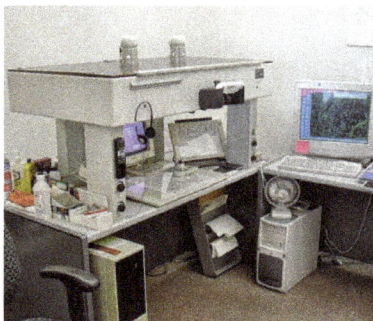

Photogrammetry

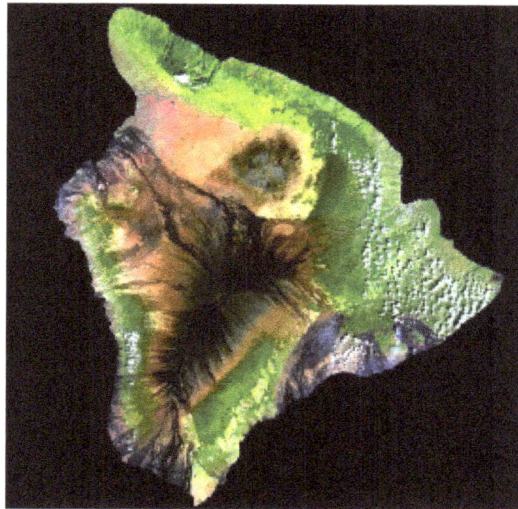

Remote Sensing

Geoinformatics Research

Research in this field is used to support global and local environmental, energy and security programs. The Geographic Information Science and Technology group of Oak Ridge National Laboratory is supported by various government departments and agencies including the United States Department of Energy. It is currently the only group in the United States Department of Energy National Laboratory System to focus on advanced theory and application research in this field. There are also a lot of interdiscipline research involved in geoinformatics fields including computer science, information technology, software engineering, biogeography, geography, conservation, architecture, spatial analysis and reinformacement learning.

Applications

Many fields benefit from geoinformatics, including urban planning and land use management, in-car navigation systems, virtual globes, public health, local and national gazetteer management, environmental modeling and analysis, military, transport network planning and management, agriculture, meteorology and climate change, oceanography and coupled ocean and atmosphere modelling, business location planning, architecture and archeological reconstruction, telecommunications, criminology and crime simulation, aviation, biodiversity conservation and maritime transport. The importance of the spatial dimension in assessing, monitoring and modelling various issues and problems related to sustainable management of natural resources is recognized all over the world. Geoinformatics becomes very important technology to decision-makers across a wide range of disciplines, industries, commercial sector, environmental agencies, local and national government, research, and academia, national survey and mapping organisations, International organisations, United Nations, emergency services, public health and epidemiology, crime mapping, transportation and infrastructure, information technology industries, GIS consulting firms, environmental management agencies), tourist industry, utility companies, market analysis and e-commerce, mineral exploration, etc. Many government and non government agencies started to use spatial data for managing their day-to-day activities.

Atmospheric Chemistry

Atmospheric chemistry is a branch of atmospheric science in which the chemistry of the Earth's atmosphere and that of other planets is studied. It is a multidisciplinary approach of research and draws on environmental chemistry, physics, meteorology, computer modeling, oceanography, geology and volcanology and other disciplines. Research is increasingly connected with other arenas of study such as climatology.

The composition and chemistry of the Earth's atmosphere is of importance for several reasons, but primarily because of the interactions between the atmosphere and living organisms. The composition of the Earth's atmosphere changes as result of natural processes such as volcano emissions, lightning and bombardment by solar particles from corona. It has also been changed by human activity and some of these changes are harmful to human health, crops and ecosystems. Examples of problems which have been addressed by atmospheric chemistry include acid rain, ozone depletion, photochemical smog, greenhouse gases and global warming. Atmospheric chemists seek to understand the causes of these problems, and by obtaining a theoretical understanding of them, allow possible solutions to be tested and the effects of changes in government policy evaluated.

Atmospheric Composition

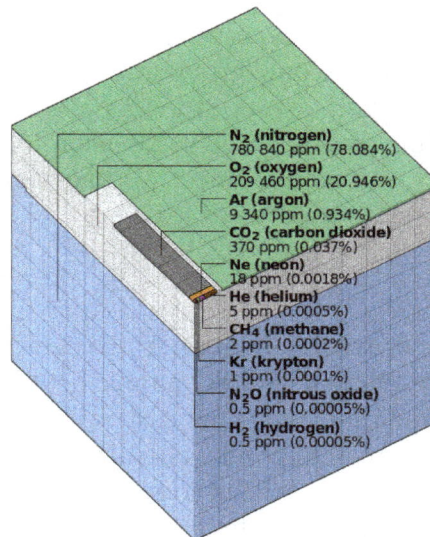

N$_2$ (nitrogen)
780 840 ppm (78.084%)
O$_2$ (oxygen)
209 460 ppm (20.946%)
Ar (argon)
9 340 ppm (0.934%)
CO$_2$ (carbon dioxide)
370 ppm (0.037%)
Ne (neon)
18 ppm (0.0018%)
He (helium)
5 ppm (0.0005%)
CH$_4$ (methane)
2 ppm (0.0002%)
Kr (krypton)
1 ppm (0.0001%)
N$_2$O (nitrous oxide)
0.5 ppm (0.00005%)
H$_2$ (hydrogen)
0.5 ppm (0.00005%)

Visualisation of composition by volume of Earth's atmosphere. Water vapour is not included as it is highly variable. Each tiny cube (such as the one representing krypton) has one millionth of the volume of the entire block. Data is from NASA Langley.

Average composition of dry atmosphere (mole fractions)		
Gas	**per NASA**	
Nitrogen, N$_2$	78.084%	
Oxygen, O$_2$	20.946%	
Minor constituents (mole fractions in ppm)		
Argon, Ar	9340	
Carbon dioxide, CO$_2$	400	

Neon, Ne	18.18	
Helium, He	5.24	
Methane, CH_4	1.7	
Krypton, Kr	1.14	
Hydrogen, H_2	0.55	
Nitrous oxide, N_2O	0.5	
Xenon, Xe	0.09	
Nitrogen dioxide, NO_2	0.02	
Water		
Water vapour	Highly variable; typically makes up about 1%	

Notes: the concentration of CO2 and CH4 vary by season and location. The mean molecular mass of air is 28.97 g/mol. Ozone (O3) is not included due to its high variability.

History

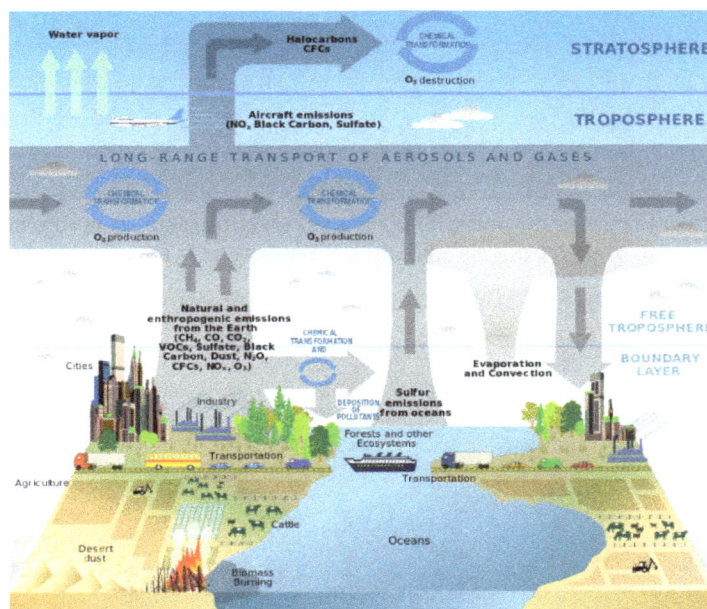

Schematic of chemical and transport processes related to atmospheric composition.

The ancient Greeks regarded air as one of the four elements, but the first scientific studies of atmospheric composition began in the 18th century. Chemists such as Joseph Priestley, Antoine Lavoisier and Henry Cavendish made the first measurements of the composition of the atmosphere.

In the late 19th and early 20th centuries interest shifted towards trace constituents with very small concentrations. One particularly important discovery for atmospheric chemistry was the discovery of ozone by Christian Friedrich Schönbein in 1840.

In the 20th century atmospheric science moved on from studying the composition of air to a consideration of how the concentrations of trace gases in the atmosphere have changed over time and the chemical processes which create and destroy compounds in the air. Two particularly important examples of this were the explanation by Sydney Chapman and Gordon Dobson of how the ozone layer is created and maintained, and the explanation of photochemical smog by Arie Jan Haagen-Smit. Further studies on ozone issues led to the 1995 Nobel Prize in Chemistry award shared between Paul Crutzen, Mario Molina and Frank Sherwood Rowland.

In the 21st century the focus is now shifting again. Atmospheric chemistry is increasingly studied as one part of the Earth system. Instead of concentrating on atmospheric chemistry in isolation the focus is now on seeing it as one part of a single system with the rest of the atmosphere, biosphere and geosphere. An especially important driver for this is the links between chemistry and climate such as the effects of changing climate on the recovery of the ozone hole and vice versa but also interaction of the composition of the atmosphere with the oceans and terrestrial ecosystems.

Carbon dioxide in Earth's atmosphere if *half of* global-warming emissions are *not absorbed*.
(NASA simulation; 9 November 2015)

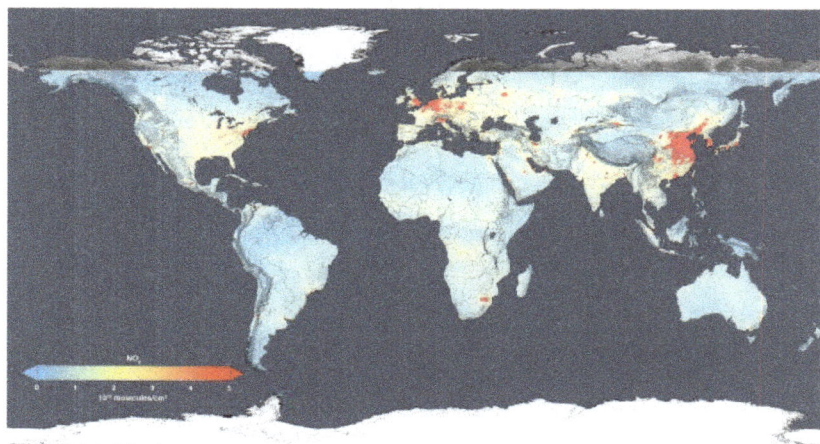

Nitrogen dioxide 2014 - global air quality levels
(released 14 December 2015).

Methodology

Observations, lab measurements and modeling are the three central elements in atmospheric chemistry. Progress in atmospheric chemistry is often driven by the interactions between these components and they form an integrated whole. For example, observations may tell us that more of a chemical compound exists than previously thought possible. This will stimulate new modelling and laboratory studies which will increase our scientific understanding to a point where the observations can be explained.

Observation

Observations of atmospheric chemistry are essential to our understanding. Routine observations of chemical composition tell us about changes in atmospheric composition over time. One important example of this is the Keeling Curve - a series of measurements from 1958 to today which show a steady rise in of the concentration of carbon dioxide. Observations of atmospheric chemistry are made in observatories such as that on Mauna Loa and on mobile platforms such as aircraft (e.g. the UK's Facility for Airborne At-mospheric Measurements), ships and balloons. Observations of atmospheric composition are in-creasingly made by satellites with important instruments such as GOME and MOPITT giving a global picture of air pollution and chemistry. Surface observations have the advantage that they provide long term records at high time resolution but are limited in the vertical and horizontal space they provide observations from. Some surface based instruments e.g. LIDAR can provide concentration profiles of chemical compounds and aerosol but are still restricted in the horizontal region they can cover. Many observations are available on line in Atmospheric Chemistry Observational Databases.

Lab Measurements

Measurements made in the laboratory are essential to our understanding of the sources and sinks of pollutants and naturally occurring compounds. Lab studies tell us which gases react with each other and how fast they react. Measurements of interest include reactions in the gas phase, on surfaces and in water. Also of high importance is photochemistry which quantifies how quickly molecules are split apart by sunlight and what the products are plus thermodynamic data such as Henry's law coefficients.

Modeling

In order to synthesise and test theoretical understanding of atmospheric chemistry, computer models (such as chemical transport models) are used. Numerical models solve the differential equations governing the concentrations of chemicals in the atmosphere. They can be very simple or very complicated. One common trade off in numerical models is between the number of chemical compounds and chemical reactions modelled versus the representation of transport and mixing in the atmosphere. For example, a box model might include hundreds or even thousands of chemical reactions but will only have a very crude representation of mixing in the atmosphere. In contrast, 3D models represent many of the physical processes of the atmosphere but due to constraints on computer resources will have far fewer chemical reactions and compounds. Models can be used to interpret observations, test understanding of chemical reactions and predict future concentrations of chemical compounds in the atmosphere. One important current trend is for atmospheric chemistry modules to become one part of earth system models in which the links between climate, atmospheric composition and the biosphere can be studied.

Some models are constructed by automatic code generators (e.g. Autochem or KPP). In this approach a set of constituents are chosen and the automatic code generator will then select the reactions involving those constituents from a set of reaction databases. Once the reactions have been chosen the ordinary differential equations (ODE) that describe their time evolution can be automatically constructed.

Atmospheric physics

Atmospheric physics is the application of physics to the study of the atmosphere. Atmospheric physicists attempt to model Earth's atmosphere and the atmospheres of the other planets using fluid flow equations, chemical models, radiation budget, and energy transfer processes in the atmosphere (as well as how these tie into other systems such as the oceans). In order to model weather systems, atmospheric physicists employ elements of scattering theory, wave propagation models, cloud physics, statistical mechanics and spatial statistics which are highly mathematical and related to physics. It has close links to meteorology and climatology and also covers the design and construction of instruments for studying the atmosphere and the interpretation of the data they provide, including remote sensing instruments. At the dawn of the space age and the introduction of sounding rockets, aeronomy became a subdiscipline concerning the upper layers of the atmosphere, where dissociation and ionization are important.

Remote Sensing

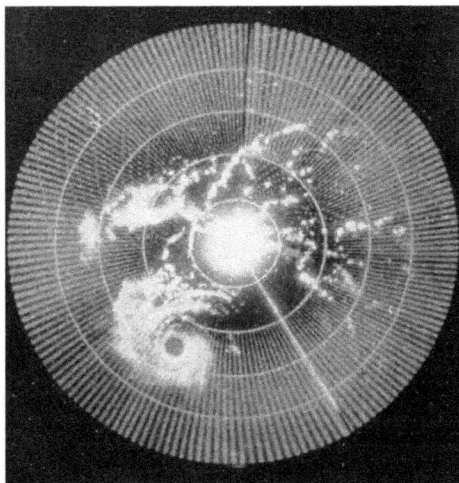

Brightness can indicate reflectivity as in this 1960 weather radar image (of Hurricane Abby). The radar's frequency, pulse form, and antenna largely determine what it can observe.

Remote sensing is the small or large-scale acquisition of information of an object or phenomenon, by the use of either recording or real-time sensing device(s) that is not in physical or intimate contact with the object (such as by way of aircraft, spacecraft, satellite, buoy, or ship). In practice, remote sensing is the stand-off collection through the use of a variety of devices for gathering information on a given object or area which gives more information than sensors at individual sites might convey. Thus, Earth observation or weather satellite collection platforms, ocean and atmospheric observing weather buoy platforms, monitoring of a pregnancy via ultrasound, Magnetic Resonance Imaging (MRI), Positron Emission Tomography (PET), and space probes are all examples of remote sensing. In modern usage, the term generally refers to the use of imaging sensor technologies including but not limited to the use of instruments aboard aircraft and spacecraft, and is distinct from other imaging-related fields such as medical imaging.

There are two kinds of remote sensing. Passive sensors detect natural radiation that is emitted or reflected by the object or surrounding area being observed. Reflected sunlight is the most common source of radiation measured by passive sensors. Examples of passive remote sensors include film photography, infra-red, charge-coupled devices, and radiometers. Active collection, on the other hand, emits energy in order to scan objects and areas whereupon a sensor then detects and measures the radiation that is reflected or backscattered from the target. radar, lidar, and SODAR are examples of active remote sensing techniques used in atmospheric physics where the time delay between emission and return is measured, establishing the location, height, speed and direction of an object.

Remote sensing makes it possible to collect data on dangerous or inaccessible areas. Remote sensing applications include monitoring deforestation in areas such as the Amazon Basin, the effects of climate change on glaciers and Arctic and Antarctic regions, and depth sounding of coastal and ocean depths. Military collection during the cold war made use of stand-off collection of data about dangerous border areas. Remote sensing also replaces costly and slow data collection on the ground, ensuring in the process that areas or objects are not disturbed.

Orbital platforms collect and transmit data from different parts of the electromagnetic spectrum, which in conjunction with larger scale aerial or ground-based sensing and analysis, provides researchers with enough information to monitor trends such as El Niño and other natural long and short term phenomena. Other uses include different areas of the earth sciences such as natural resource management, agricultural fields such as land usage and conservation, and national security and overhead, ground-based and stand-off collection on border areas.

Radiation

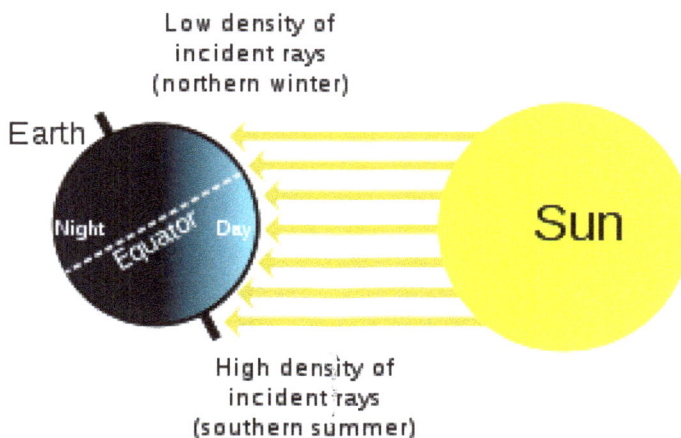

This is a diagram of the seasons. In addition to the density of incident light, the dissipation of light in the atmosphere is greater when it falls at a shallow angle.

Atmospheric physicists typically divide radiation into solar radiation (emitted by the sun) and terrestrial radiation (emitted by Earth's surface and atmosphere).

Solar radiation contains variety of wavelengths. Visible light has wavelengths between 0.4 and

0.7 micrometers. Shorter wavelengths are known as the ultraviolet (UV) part of the spectrum, while longer wavelengths are grouped into the infrared portion of the spectrum. Ozone is most effective in absorbing radiation around 0.25 micrometers, where UV-c rays lie in the spectrum. This increases the temperature of the nearby stratosphere. Snow reflects 88% of UV rays, while sand reflects 12%, and water reflects only 4% of incoming UV radiation. The more glancing the angle is between the atmosphere and the sun's rays, the more likely that energy will be reflected or absorbed by the atmosphere.

Terrestrial radiation is emitted at much longer wavelengths than solar radiation. This is because Earth is much colder than the sun. Radiation is emitted by Earth across a range of wavelengths, as formalized in Planck's law. The wavelength of maximum energy is around 10 micrometers.

Cloud Physics

Cloud physics is the study of the physical processes that lead to the formation, growth and precipitation of clouds. Clouds are composed of microscopic droplets of water (warm clouds), tiny crystals of ice, or both (mixed phase clouds). Under suitable conditions, the droplets combine to form precipitation, where they may fall to the earth. The precise mechanics of how a cloud forms and grows is not completely understood, but scientists have developed theories explaining the structure of clouds by studying the microphysics of individual droplets. Advances in radar and satellite technology have also allowed the precise study of clouds on a large scale.

Atmospheric Electricity

Cloud to ground Lightning in the global atmospheric electrical circuit.

Atmospheric electricity is the regular diurnal variations of the Earth's atmospheric electromagnetic network (or, more broadly, any planet's electrical system in its layer of gases). The Earth's surface, the ionosphere, and the atmosphere is known as the global atmospheric electrical circuit.

Lightning discharges 30,000 amperes, at up to 100 million volts, and emits light, radio waves, x-rays and even gamma rays. Plasma temperatures in lightning can approach 28,000 kelvins and electron densities may exceed $1024/m^3$.

Atmospheric Tide

The largest-amplitude atmospheric tides are mostly generated in the troposphere and stratosphere when the atmosphere is periodically heated as water vapour and ozone absorb solar radiation during the day. The tides generated are then able to propagate away from these source regions and ascend into the mesosphere and thermosphere. Atmospheric tides can be measured as regular fluctuations in wind, temperature, density and pressure. Although atmospheric tides share much in common with ocean tides they have two key distinguishing features:

i) Atmospheric tides are primarily excited by the Sun's heating of the atmosphere whereas ocean tides are primarily excited by the Moon's gravitational field. This means that most atmospheric tides have periods of oscillation related to the 24-hour length of the solar day whereas ocean tides have longer periods of oscillation related to the lunar day (time between successive lunar transits) of about 24 hours 51 minutes.

ii) Atmospheric tides propagate in an atmosphere where density varies significantly with height. A consequence of this is that their amplitudes naturally increase exponentially as the tide ascends into progressively more rarefied regions of the atmosphere. In contrast, the density of the oceans varies only slightly with depth and so there the tides do not necessarily vary in amplitude with depth.

Note that although solar heating is responsible for the largest-amplitude atmospheric tides, the gravitational fields of the Sun and Moon also raise tides in the atmosphere, with the lunar gravitational atmospheric tidal effect being significantly greater than its solar counterpart.

At ground level, atmospheric tides can be detected as regular but small oscillations in surface pressure with periods of 24 and 12 hours. Daily pressure maxima occur at 10 a.m. and 10 p.m. local time, while minima occur at 4 a.m. and 4 p.m. local time. The absolute maximum occurs at 10 a.m. while the absolute minimum occurs at 4 p.m. However, at greater heights the amplitudes of the tides can become very large. In the mesosphere (heights of ~ 50 – 100 km) atmospheric tides can reach amplitudes of more than 50 m/s and are often the most significant part of the motion of the atmosphere.

Aeronomy

Aeronomy is the science of the upper region of the atmosphere, where dissociation and ionization are important. The term aeronomy was introduced by Sydney Chapman in 1960. Today, the term also includes the science of the corresponding regions of the atmospheres of other planets. Research in aeronomy requires access to balloons, satellites, and sounding rockets which provide valuable data about this region of the atmosphere. Atmospheric tides play an important role in interacting with both the lower and upper atmosphere. Amongst the phenomena studied are upper-atmospheric lightning discharges, such as luminous events called red sprites, sprite halos, blue jets, and elves.

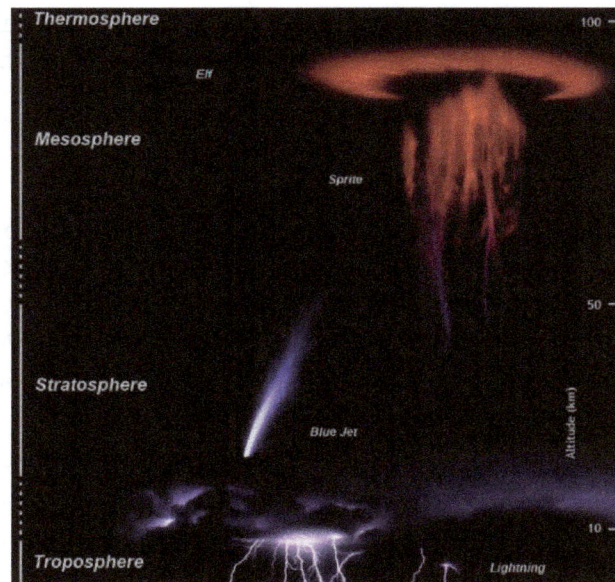

Representation of upper-atmospheric lightning and electrical-discharge phenomena

Centers of Research

In the UK, atmospheric studies are underpinned by the Met Office, the Natural Environment Research Council and the Science and Technology Facilities Council. Divisions of the U.S. National Oceanic and Atmospheric Administration (NOAA) oversee research projects and weather modeling involving atmospheric physics. The US National Astronomy and Ionosphere Center also carries out studies of the high atmosphere. In Belgium, the Belgian Institute for Space Aeronomy studies the atmosphere and outer space.

References

- Bozorgnia, Yousef; Bertero, Vitelmo V. (2004). Earthquake Engineering: From Engineering Seismology to Performance-Based Engineering. CRC Press. ISBN 978-0-8493-1439-1.

- Davies, Geoffrey F. (2001). Dynamic Earth: Plates, Plumes and Mantle Convection. Cambridge University Press. ISBN 0-521-59067-1.

- Eratosthenes (2010). Eratosthenes' "Geography". Fragments collected and translated, with commentary and additional material by Duane W. Roller. Princeton University Press. ISBN 978-0-691-14267-8.

- Merrill, Ronald T.; McElhinny, Michael W.; McFadden, Phillip L. (1998). The Magnetic Field of the Earth: Paleomagnetism, the Core, and the Deep Mantle. International Geophysics Series. 63. Academic Press. ISBN 978-0124912458.

- Poirier, Jean-Paul (2000). Introduction to the Physics of the Earth's Interior. Cambridge Topics in Mineral Physics & Chemistry. Cambridge University Press. ISBN 0-521-66313-X.

- Sheriff, Robert E. (1991). "Geophysics". Encyclopedic Dictionary of Exploration Geophysics (3rd ed.). Society of Exploration. ISBN 978-1-56080-018-7.

- Telford, William Murray; Geldart, L. P.; Sheriff, Robert E. (1990). Applied geophysics. Cambridge University Press. ISBN 978-0-521-33938-4.

- Reinhardt, Carsten (2008). Chemical Sciences in the 20th Century: Bridging Boundaries. John Wiley & Sons. p. 161. ISBN 3-527-30271-9.

- Brian Mason (1992). Victor Moritz Goldschmidt: Father of Modern Geochemistry (Geochemical Society). ISBN 0-941809-03-X

- Dilek, Yildirim; Harald Furnes; Karlis Muehlenbachs (2008). Links Between Geological Processes, Microbial Activities & Evolution of Life. Springer. p. v. ISBN 1-4020-8305-X.

- Knoll, Andrew H.; Canfield, Professor Don E.; Konhauser, Kurt O. (2012-03-30). Fundamentals of Geobiology. John Wiley & Sons. ISBN 9781118280881.

- Bekker, Andrey (2014-01-01). Amils, Ricardo; Gargaud, Muriel; Quintanilla, José Cernicharo; Cleaves, Henderson James; Irvine, William M.; Pinti, Daniele; Viso, Michel, eds. Great Oxygenation Event. Springer Berlin Heidelberg. pp. 1–9. doi:10.1007/978-3-642-27833-4_1752-4. ISBN 9783642278334.

- Ehrlich, Henry Lutz; Newman, Dianne K.; Kappler, Andreas (2015-10-15). Ehrlich's Geomicrobiology, Sixth Edition. CRC Press. ISBN 9781466592414.

- L, Slonczewski, Joan; W, Foster, John (2013-10-01). Microbiology: An Evolving Science: Third International Student Edition. W. W. Norton & Company. ISBN 9780393923216.

- Knoll, Andrew H. (2015-03-22). Life on a Young Planet: The First Three Billion Years of Evolution on Earth. Princeton University Press. ISBN 9781400866045.

Earth System Science: A Comprehensive Study

Earth system science offers a holistic view into the functioning of planet Earth. It considers the movement of the planet along with the changes in the seasons as subjects that need to be studied together rather than separately. It also considers human activity such as waste disposal and emissions to be objects of study along with celestial activity.

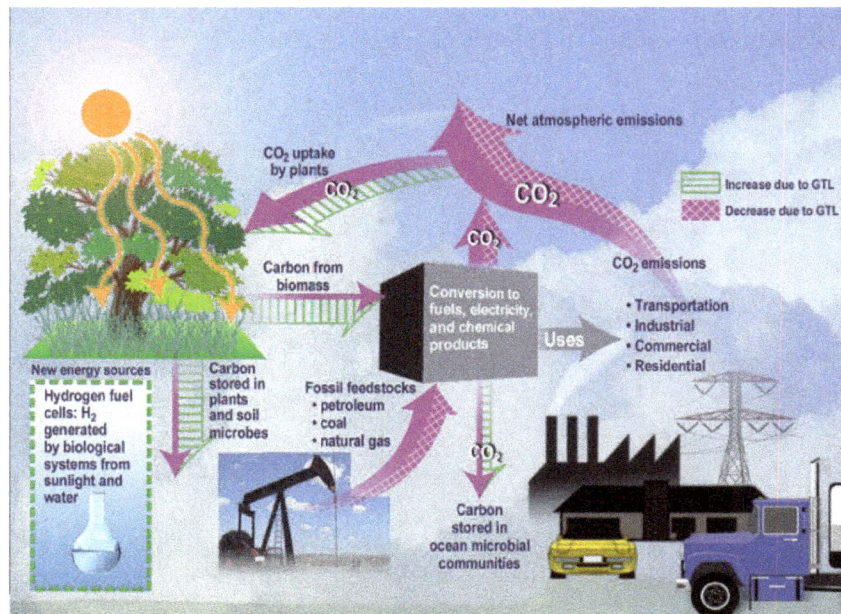

An ecological analysis of CO_2
in an ecosystem. As systems biology, systems ecology seeks a holistic view of the interactions and transactions within and between biological and ecological systems.

Earth system science (ESS) is the application of systems science to the Earth sciences. In particular, it considers interactions between the Earth's "spheres"—atmosphere, hydrosphere, cryosphere, geosphere, pedosphere, biosphere, and, even, the magnetosphere—as well as the impact of human societies on these components. At its broadest scale, Earth system science brings together researchers across both the natural and social sciences, from fields including ecology, economics, geology, glaciology, meteorology, oceanography, paleontology, sociology, and space science. Like the broader subject of systems science, Earth system science assumes a holistic view of the dynamic interaction between the Earth's spheres and their many constituent subsystems, the resulting organization and time evolution of these systems, and their stability or instability. Subsets of Earth system science include systems geology and systems ecology, and many aspects of Earth system science are fundamental to the subjects of physical geography and climate science.

Definition

The Science Education Resource Center, Carleton College, offers the following description: "Earth system science embraces chemistry, physics, biology, mathematics and applied sciences in transcending disciplinary boundaries to treat the Earth as an integrated system. It seeks a deeper understanding of the physical, chemical, biological and human interactions that determine the past, current and future states of the Earth. Earth system science provides a physical basis for understanding the world in which we live and upon which humankind seeks to achieve sustainability".

Origins

For millennia, humans have speculated how the physical and living elements on the surface of the Earth combine, with gods and goddesses frequently posited to embody specific elements. The notion that the Earth, itself, is alive was a regular theme of Greek philosophy and religion. Early scientific interpretations of the Earth system began in the field of geology, initially in the Middle East and China, and largely focused on aspects such as the age of the Earth and the large-scale processes involved in mountain and ocean formation. As geology developed as a science, understanding of the interplay of different facets of the Earth system increased, leading to the inclusion of factors such as the Earth's interior, planetary geology and living systems.

In many respects, the foundational concepts of Earth system science can be seen in the holistic interpretations of nature promoted by the 19th century geographer Alexander von Humboldt. In the 20th century, Vladimir Vernadsky (1863-1945) saw the functioning of the biosphere as a geological force generating a dynamic disequilibrium, which in turn promoted the diversity of life. In the mid-1960s, James Lovelock first postulated a regulatory role for the biosphere in feedback mechanisms within the Earth system. Initially named the "Earth Feedback hypothesis", Lovelock later renamed it the Gaia hypothesis, and subsequently further developed the theory with American evolutionary theorist Lynn Margulis during the 1970s. In parallel, the field of systems science was developing across numerous other scientific fields, driven in part by the increasing availability and power of computers, and leading to the development of climate models that began to allow the detailed and interacting simulations of the Earth's weather and climate. Subsequent extension of these models has led to the development of "Earth system models" (ESMs) that include facets such as the cryosphere and the biosphere.

As an integrative field, Earth system science assumes the histories of a vast range of scientific disciplines, but as a discrete study it evolved in the 1980s, particularly at NASA, where a committee called the Earth System Science Committee was formed in 1983. The earliest reports of NASA's ESSC, *Earth System Science: Overview* (1986), and the book-length *Earth System Science: A Closer View* (1988), constitute a major landmark in the formal development of Earth system science. Early works discussing Earth system science, like these NASA reports, generally emphasized the increasing human impacts on the Earth system as a primary driver for the need of greater integration among the life and geo-sciences, making the origins of Earth system science parallel to the beginnings of global change studies and programs.

Climate Science and Earth System Science

Climatology and climate change have been central to Earth system science since its inception, as evidenced by the prominent place given to climate change in the early NASA reports discussed above. The Earth's climate system is a prime example of an emergent property of the whole planetary system which cannot be fully understood without regarding it as a single integrated entity. It is also a property of the system where human impacts have been growing rapidly in recent decades, lending immense importance to the successful development and advancement of Earth system science research. As just one example of the centrality of climatology to the field, leading American climatologist Michael E. Mann is the Director of one of the earliest centers for Earth system science research, the Earth System Science Center at Pennsylvania State University, and its mission statement reads, "the Earth System Science Center (ESSC) maintains a mission to describe, model, and understand the Earth's climate system".

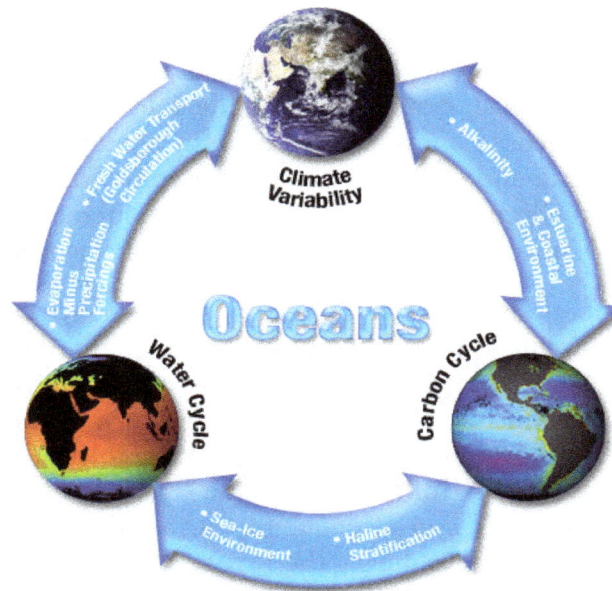

The dynamic interaction of the Earth's oceans, climatological, geochemical systems.

Relationship to the Gaia Hypothesis

The Gaia hypothesis posits that living systems interact with physical components of the Earth system to form a self-regulating whole that maintains conditions that are favourable for life. Developed initially by James Lovelock, the hypothesis attempts to account for key features of the Earth system, including the long period (several billion years) of relatively favourable climatic conditions against a backdrop of steadily increasing solar radiation. Consequently, the Gaia hypothesis has important implications for Earth system science, as noted by NASA's Director for Planetary Science, James Green, in October 2010: "Dr. Lovelock and Dr. Margulis played a key role in the origins of what we now know as Earth system science".

Although the Gaia hypothesis and Earth system science take an interdisciplinary approach to studying systems operations on a planetary-scale, they are not synonymous with one another. A number of potential Gaian feedback mechanisms have been proposed — such as the CLAW hypothesis — but the hypothesis does not have universal support within the scientific community, though it remains an active research topic.

Education

Earth system science can be studied at a postgraduate level at some universities, with notable programs at such institutions as the University of Pennsylvania, Stanford University, and the University of California, Santa Cruz. In general education, the American Geophysical Union, in cooperation with the Keck Geology Consortium and with support from five divisions within the National Science Foundation, convened a workshop in 1996, "to define common educational goals among all disciplines in the Earth sciences". In its report, participants noted that, "The fields that make up the Earth and space sciences are currently undergoing a major advancement that promotes understanding the Earth as a number of interrelated systems". Recognizing the rise of this systems approach, the workshop report recommended that an Earth system science curriculum be developed with support from the National Science Foundation. In 2000, the Earth System Science Education Alliance was begun, and currently includes the participation of 40+ institutions, with over 3,000 teachers having completed an ESSEA course as of fall 2009".

References

- Jacobson, Michael; et al. (2000). Earth System Science, From Biogeochemical Cycles to Global Changes (2nd ed.). London: Elsevier Academic Press. ISBN 978-0123793706. Retrieved 7 September 2015.

- Asimov, M. S.; Bosworth, Clifford Edmund (eds.). The Age of Achievement: A.D. 750 to the End of the Fifteenth Century : The Achievements. History of civilizations of Central Asia. pp. 211–214. ISBN 978-92-3-102719-2.

- Beerling, David (2007). The Emerald Planet: How plants changed Earth's history. Oxford: Oxford University Press. ISBN 978-0-19-280602-4.

- Tyrrell, Toby (2013), On Gaia: A Critical Investigation of the Relationship between Life and Earth, Princeton: Princeton University Press, ISBN 9780691121581.

- Tickell, Crispin (2006). "Earth Systems Science: Are We Pushing Gaia Too Hard?". 46th Annual Bennett Lecture - University of Leicester. London: University of Leicester. Retrieved 2015-09-21.

- Schneider, Stephen; Boston, Penelope (1992). "The Gaia Hypothesis and Earth System Science" (PDF). University of Florida. MIT Press. Retrieved 2015-09-21.

- Edwards, P.N. (2010). "History of climate modelling". Wiley Interdisciplinary Reviews: Climate Change. John Wiley & Sons. 2: 128–139. doi:10.1002/wcc.95. Retrieved 2 October 2015.

- Washington, W.M.; Buja, L.; Craig, A. (2009). "The computational future for climate and Earth system models: on the path to petaflop and beyond". Phil. Trans. Roy. Soc. A. Royal Society. 367: 833–846. doi:10.1098/rsta.2008.0219. Retrieved 2 October 2015.

- Mooney, Harold; et al. (February 26, 2013). "Evolution of natural and social science interactions in global change research programs". Proceedings of the National Academy of Sciences. 110 (Supplement 1, 3665-3672): 3665–3672. doi:10.1073/pnas.1107484110. Retrieved 7 September 2015.

- NASA, 50th Anniversary Symposium: Seeking Signs of Life. "Opening Keynote - 'Exobiology in the Beginning'". livestream.com. Retrieved 7 September 2015.

Permissions

Index

www.ingramcontent.com/pod-product-compliance
Lightning Source LLC
Chambersburg PA
CBHW061248190326

41458CB00011B/3609